submarine telecoms

FORUM

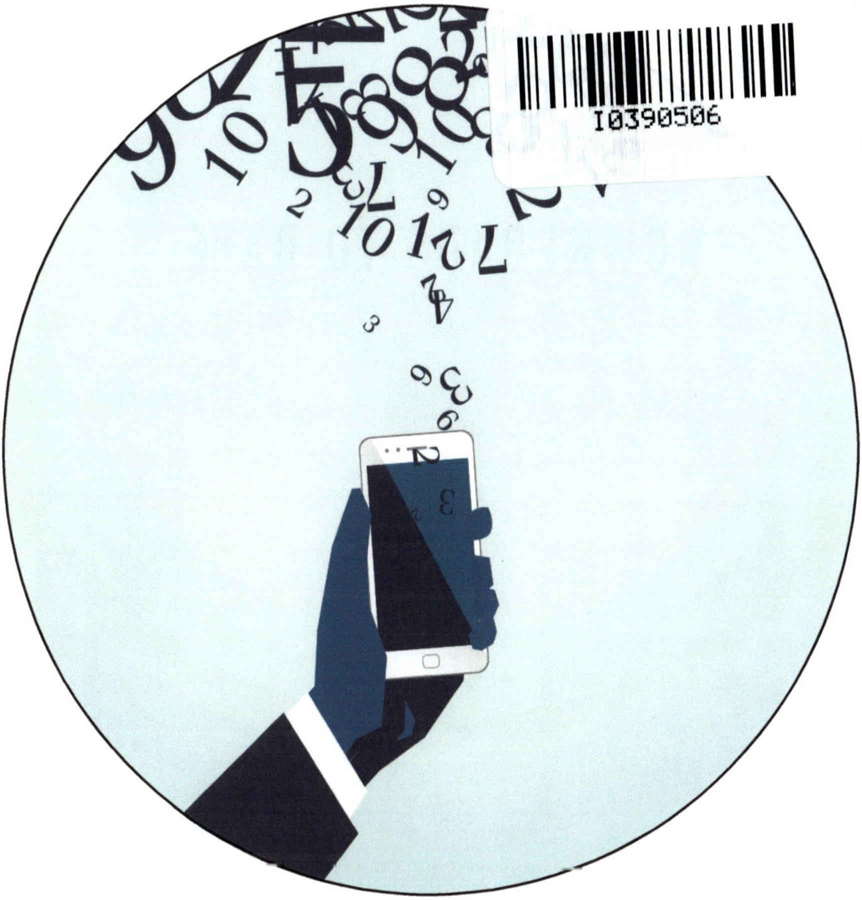

Global Outlook Edition

86

WEBSITE TRAFFIC: UNIQUE VISITS

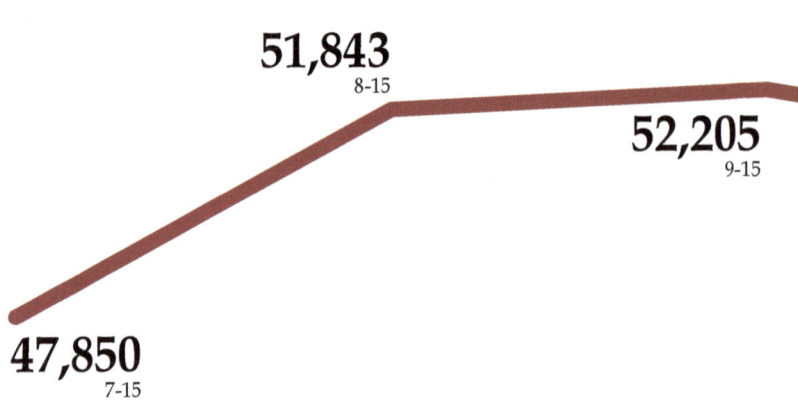

- 47,850 (7-15)
- 51,843 (8-15)
- 52,205 (9-15)

DOWNLOADS TO DATE:

47,668

Issue #83 - Released 7-15

58,659

Issue #84 - Released 9-15

54,874

Issue #85 - Released 11-15

TOTAL HITS IN 2015:
10,303,895

50,999 10-15

50,713 11-15

47,612 12-15

572,751

520,542

172,801

Issue #14 - Released 5-15

Issue #15 - Released 8-15

Issue #16 - Released 11-15

EXORDIUM
BY WAYNE NIELSEN

Welcome to Issue 86, our Global Outlook edition.

I spent the last few days mulling over what to talk about for my part in the opening submarine cable session at PTC '16.

I am normally quite happy to sit quietly in the audience and listen to others opine about our industry, and learn a few things in the process. But instead, around Halloween, the estimable Paul McCann contacted me and asked if I would be willing to participate in the session. I thought about it for a while and then that I would accept before he could revoke the invitation. Then I proceeded to forget about the commitment until sometime after Christmas.

So, the new year came and went, and I returned to the office that Monday from the end-of-year sabbatical from that day-job, and to the unwritten presentation that I had still not really started.

But a funny thing happened.

One of the benefits of being the titular SubTel Forum head is that I get to see articles in their draft stage, and, low and behold, a couple of articles from a couple of really smart guys came through the door. It wasn't that they wrote the presentation for me, but instead enhanced a thought I had been having, thus allowing me to go in the direction I wanted. And of course, accreditations abound within.

So what you might ask did I conclude as an outlook for the industry?

That, as Dr. Seuss once opined, "You're in pretty good shape for the shape you are in."

So, as always, should you be attending PTC '16, please come to the submarine cable session and our SubTel Forum booth to say hello; and of course, save me a seat at the mai tai bar!

W Nielsen

Wayne Nielsen is the Founder and Publisher of Submarine Telecoms Forum, and previously in 1991, founded and published "Soundings", a print magazine developed for then BT Marine. In 1998, he founded and published for SAIC the magazine, "Real Time", the industry's first electronic magazine. He has written a number of industry papers and articles over the years, and is the author of two published novels, Semblance of Balance (2002, 2014) and Snake Dancer's Song (2004).

 +1.703.444.2527

 wnielsen@subtelforum.com

IN THIS ISSUE....

4	Exordium *Wayne Nielsen*
8	Advertiser Index
10	News Now
18	Global Outlook *Kieran Clark*
26	The Knowledge Economy, Its Growth In Technology And Virtualisation *Derek Cassidy*
38	Is Industry Consolidation Overdue? *Jas Dhooper*
56	The Atlantic: An Infrastructure Analysis *Hubert Souisa*

68   2016: A Year Of 'Radical' Change In Subsea System Development?
     John Tibbles

88   Realizing The Value Of A Connected Africa
     Mike Last

104 Advertiser's Corner
     Kristian Nielsen

108 Coda
     Kevin G. Summers

ADVERTISER INDEX

OFS	www.ofsoptics.com	**22**
SubOptic	www.suboptic.org	**16**
WFN Strategies	www.wfnstrategies.com	**102**

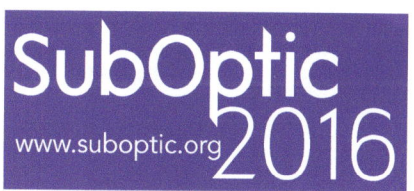

News Now

- ADB Grants USD 25 Million for Samoa-Fiji Submarine Cable

- Agartala Set to be India's Third Internet Gateway

- Alcatel-Lucent and Bluesky Pacific Group Launch New Submarine Cable System to Enhance Connectivity Across Pacific

- Alcatel-Lucent to Start Construction of ACE Undersea Cable System's Next Phase

- America-Europe Connect Submarine Cable Network Completes Marine Construction

- Brazil-Europe Submarine Cable to Start Construction in April

- Cinia Finishes Submarine Cable to Link Germany, Finland

- Fibre-Optic Operators Call for Cable Protection Law

- Google to Support Uruguay-Brazil Fibre Cable

- Huawei Marine and the WACS Consortium Go Live with Underwater Submarine Cable following 100G Upgrade

- Inaugural Workshop Held in Indonesia with the ICPC and the Coordinating Ministry for Maritime Affairs

- India's New Internet Gateway Via Cox's Bazar to Open Late January, Says Minister

- Industry Report Available Now

- Infinera Names Scott Jackson to Lead Subsea Business Group

- Liberty Global Snaps Up Cable & Wireless in £3.5bn Deal

- Local Kenyan Firm Plans Sh10bn Investment in Data Centres

- MainOne Upgrades Submarine Cable Network to 100G with Xtera

News Now

- New Subsea Cable to Take Internet Connection to Landlocked African Countries

- Nigeria-Cameroun Submarine Cable Goes Live

- Pioneer Consulting Appoints Iain Ritson as Senior Project Manager

- Poor Network Hits Telecom Services in Tripura

- Reliance Jio's AAE-I Subsea Cable Project Gets a Green Light from the Green Panel

- Rini on Solomon Islands Submarine Cable

- SEA-ME-WE 5 cable system lands in Malaysia

- Seaborn Networks Completes Total Funding of US$500 Million for Seabras-1

- Sierra Leone News: 'Govt Fully Committed to Multiple Gateways' – World Bank Manager Says
- Solomon Islands High-Speed Cable Operational by October 2016
- STF Radio: PTC'16
- Submarine Cable Networks to Grow 29 Percent Getween 2014-2016, FCC Says
- SubOptic 2016 – Keynote Speakers
- SubOptic 2016 – Register Now for This Must Attend Event.
- SubOptic 2016 – The Masterclass/Tutorial Programme
- SubOptic 2016 – Workshop on Extending the Life of Submarine Cable Systems
- TE SubCom Boost Cable Manufacturing Capacity

News Now

➤ Tektronix and TE SubCom Partner on Optical Modulation Analysis Measurements

➤ Telecom Italia Sparkle and Cyta Announce the Agreement for the New "KIMONAS" Subsea Optical Fibre Cable Subsystem

➤ The Clock Is Ticking – Comment Deadlines Set for FCC's Submarine Cable Network Outage Reporting Rulemaking

➤ This Week in Submarine Telecoms January 4-8

➤ This Week in Submarine Telecoms November 16-20

➤ This Week in Submarine Telecoms November 2-6

➤ This Week in Submarine Telecoms November 30 – December 4

➤ This Week in Submarine Telecoms November 9-13

- TI Sparkle announces US$300 million long term investment in Seabras-1 cable connecting Brazil and US

- Vocus Communication and NextGen Networks in $198m Deal for New Submarine Cable

- WFN Strategies Receives 2015 American Leadership Award

- WFN Strategies to Present and Exhibit at PTC '16

- Work Starts on ACE cable

- Zayo to Acquire Viatel

SubOptic 2016
www.suboptic.org

Emerging Subsea Networks

To see the latest information about SubOptic 2016 go to our website – www.suboptic.org

The world's expanding treasure

Dubai
18th-21st April 2016

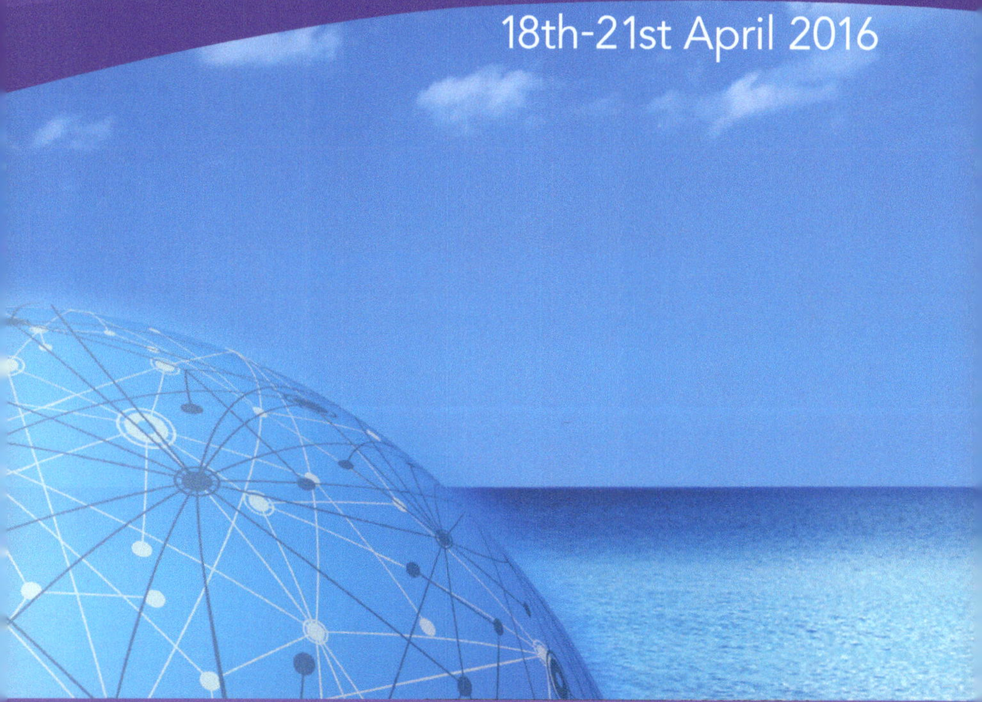

Celebrating **30** *years of SubOptic*

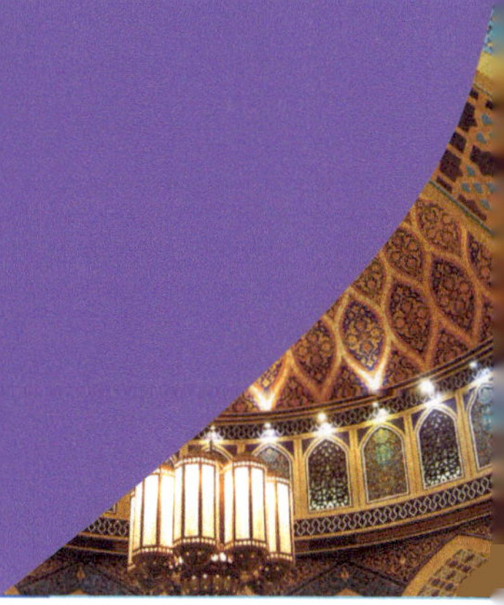

Hosted by

GLOBAL OUTLOOK
BY KIERAN CLARK

The submarine fiber industry has hit something of a rough patch in recent times. In 2013, the industry had plans for 19 systems to go live in 2014, and 29 to go live in 2015. The actual implementation rate shows a pretty bleak picture, with only 7 systems making it into service for all of 2014, and a mere 6 systems going live in 2015. Many factors contributed to this low ready-for-service rate including, but not limited to, an overall weaker global economy and the prevalence of cheap system upgrades.

However, while the past looks gloomy, the next two years should breathe new life into the submarine fiber market. The industry as a whole should be very busy, as indicated by a relatively high contract-in-force rate for planned systems.

Welcome to SubTel Forum's annual Global Outlook issue. This month, we'll take a brief look at how the industry performed around the world last year, and look ahead to 2016. The data used in this article is obtained from the public domain and is tracked by the ever-evolving SubTel Forum database, where products like the Almanac, Cable Map, and Industry Report find their roots.

Our last Global Outlook edition reported 21 systems planned to be ready for

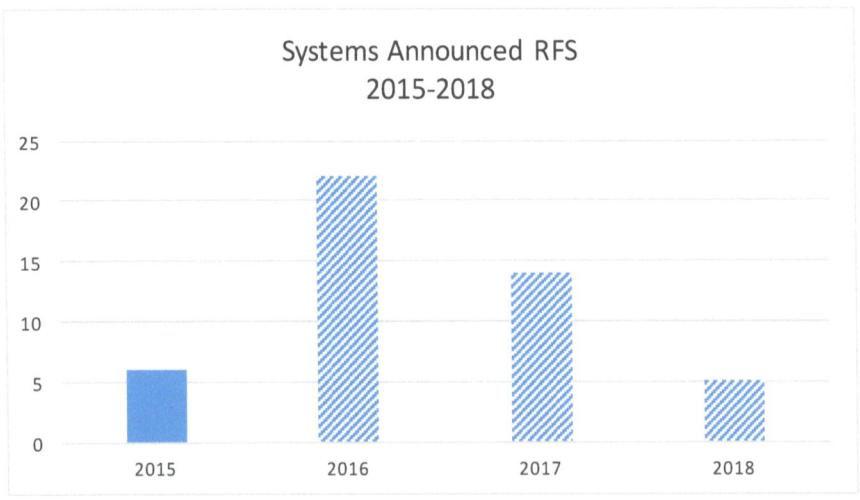

service in 2015, and 14 systems for 2016. One year later, these numbers look drastically different. Only 6 systems went live in 2015, while 2016 saw a significant increase to 22 planned systems. A couple of systems planned for 2015 simply slipped their RFS date to 2016, but more than half died outright. The outlook for 2016 is already brighter, with several systems on track to be ready for service in the first quarter of this year, and a much lower attrition rate compared to years past.

Naturally, with a reduction in the number of systems that were supposed to enter service for 2015, a decrease in the total kilometers of cable added is expected. At the start of 2015, over 156,000 kilometers of cable were planned to be added. Unfortunately, when all was said and done, a mere 15,800 kilometers of cable were laid across the world for the entirety of 2015. This huge decline is a result of the uncertainty the industry has been experiencing around the world for the past several years, causing projects to go into limbo or outright dying due to their business cases completely falling apart. Overall, 2015 missed its planned kilometers mark by over 140,000.

On the other hand, planed kilometers for 2016 have done nothing but increase

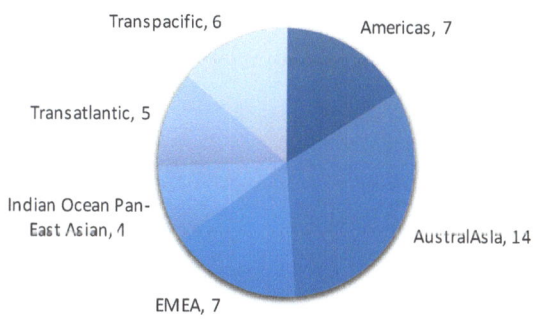

over the past two years. Desire for lower latency systems in developed regions and interest in brand new routes – particularly in the South Atlantic – have sparked something of a resurgence. This time last year, 2016 was projected to add about 120,000 kilometers of cable globally. At the start of 2016, that number has climbed to just over 130,000 kilometers. Additionally, while plans for 2017 and 2018 are understandably still being ironed out, the industry can expect up to 70,000 and 50,000 kilometers of new cable, respectively.

As has been the trend for several years running, the bulk of new system activity is occurring in the Pacific Ocean. Growth along Transpacific routes has been spurred by a significantly increased desire for lower latency between Asia, Australia, and the United States. AustralAsia continues to experience explosive growth driven by emerging markets in the South Pacific, despite the current economic woes of China. The EMEA and Indian Ocean Pan-East Asian regions continue to see muted growth, largely due to increasing political and economic instability in the Middle East and the saturation of African telecommunications markets. Transatlantic growth is primarily being driven by a desire to connect

From shore to shore . . .
OFS' complete portfolio of ocean fiber solutions enables industry-leading performance through high signal power and low-loss.

A Furukawa Company

To learn more, ask your cabler about OFS or visit www.ofsoptics.com

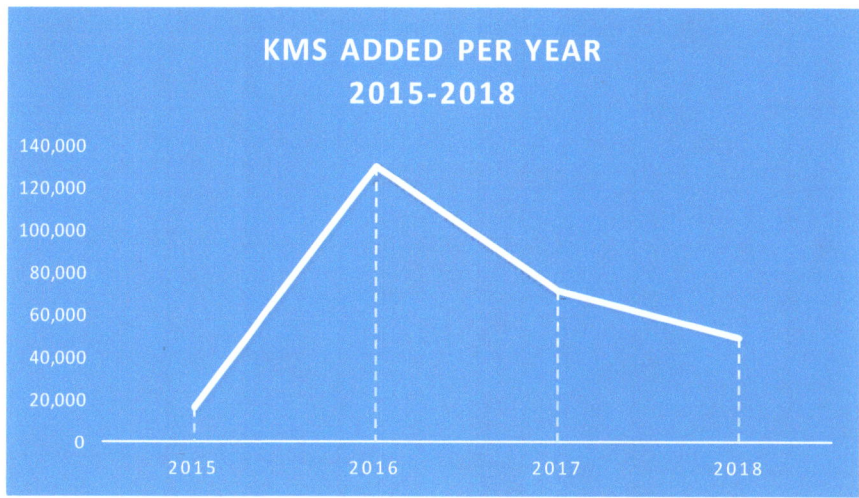

South America directly to Europe and Africa, alongside new interest in lower latency routes between the United States and Europe. The Americas region is characterized by replacing older systems in the Caribbean and increased demand for connectivity between the United States and Brazil.

Nearly every region in the world has seen an uptick in planned activity for the next few years. Now more than ever, new systems in developed regions are focusing on latency, rather than bandwidth. This has

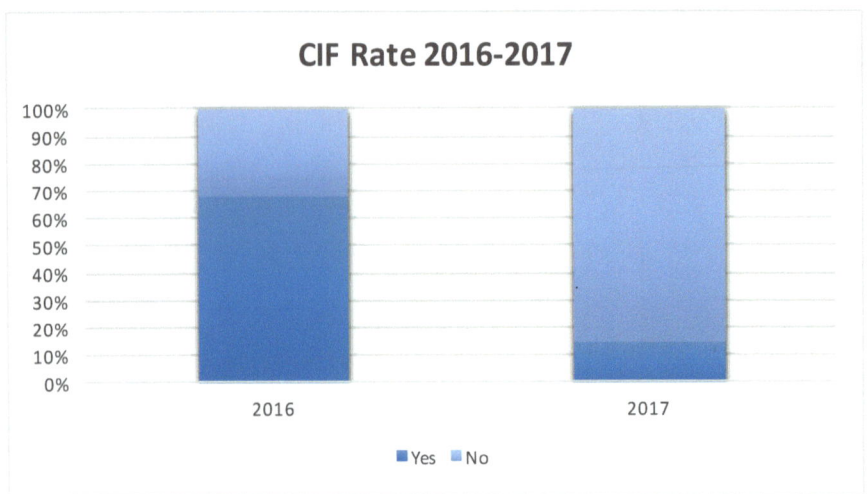

allowed for growth in regions that haven't seen a significant new system in years, such as the Transatlantic.

Of course, the first real test of a system's viability is whether or not it is contract-in-force. As of the new year, 47% of all systems planned for 2016 and 2017 are contract-in-force. Considering the dismally low CIF rate of 24% for our Regional Outlook issue back in July – which included systems still planned for 2015 – this is an incredibly encouraging number. Looking at 2016 by itself

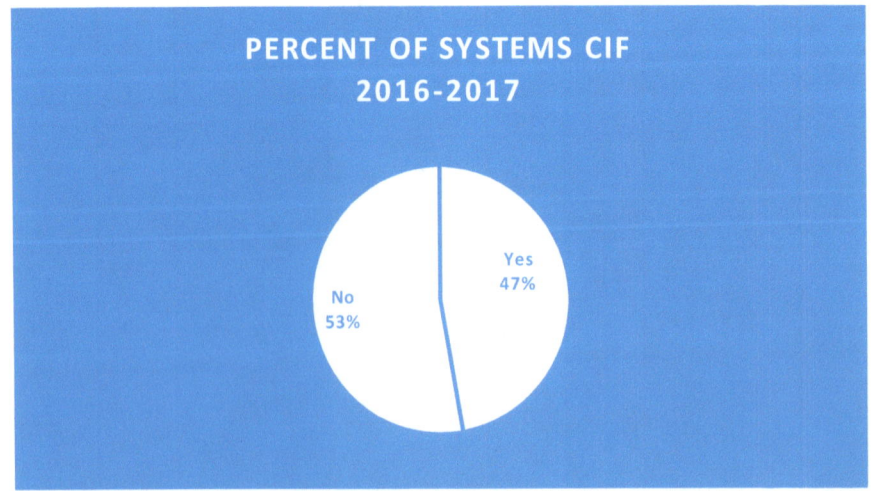

shows a CIF rate of 70%, an even more promising number. Considering this high CIF rate has been achieved by the first week of January, it's pretty safe to say 2016 looks to be a much needed boon for the submarine fiber industry.

Overall, the industry is potentially on the cusp of a major resurgence. The last few years have been lean, scraping by largely on the back of a dwindling upgrade market and new routes to small nations in the Pacific. All bets are off for the next two years though, as the quest for lower latency becomes the main focus in developed regions and critical new routes are explored between major emerging economies. Everyone in the industry will be busy for the near future, with plenty of projects to go around.

Kieran Clark is an Analyst for Submarine Telecoms Forum. He joined the company in 2013 as a Broadcast Technician to provide support for live event video streaming. In 2014, Kieran was promoted to Analyst and is currently responsible for the research and maintenance that supports the SubTel Forum International Submarine Cable Database; his analysis is featured in almost the entire array of SubTel Forum publications. He has 4+ years of live production experience and has worked alongside some of the premier organizations in video web streaming.

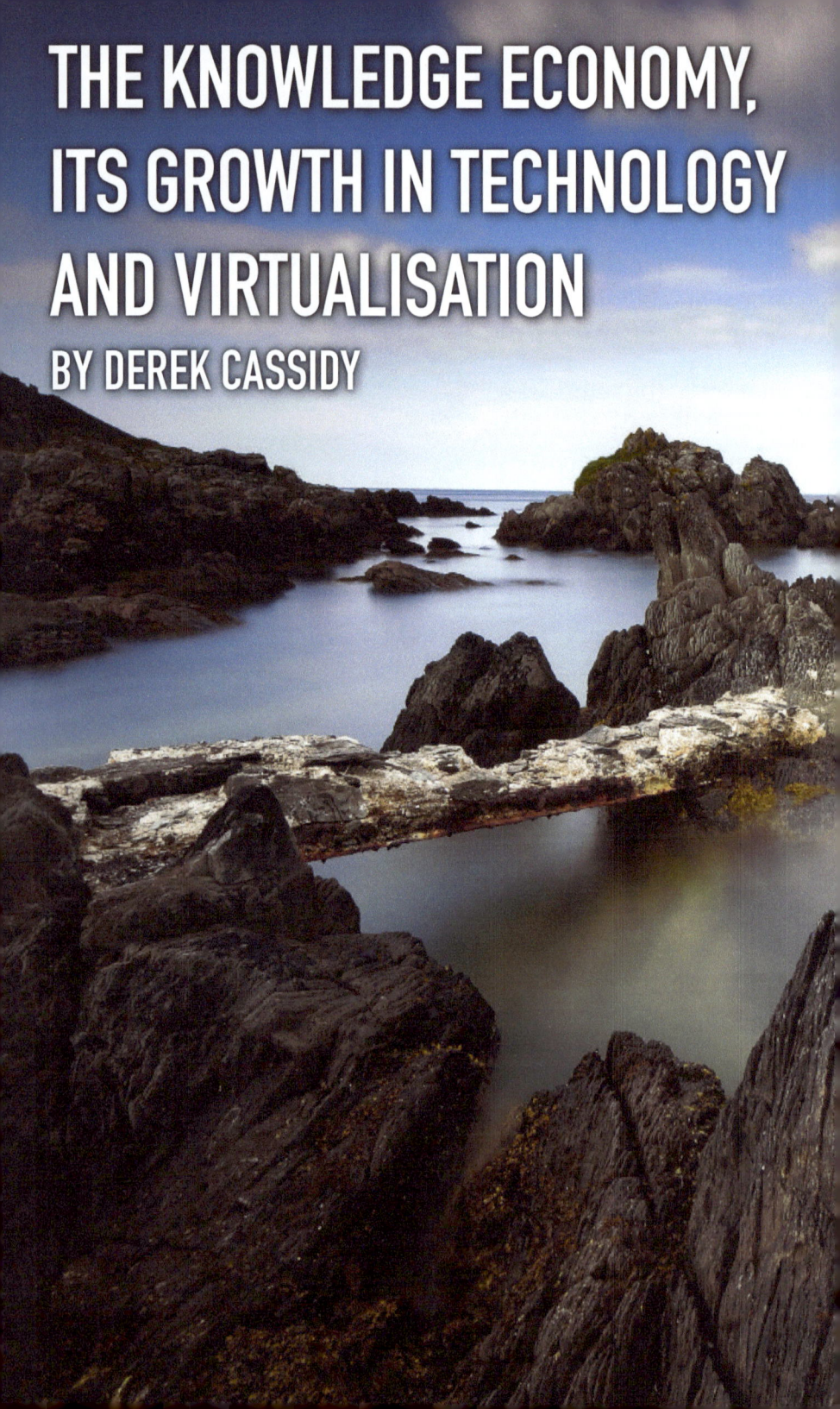

THE KNOWLEDGE ECONOMY, ITS GROWTH IN TECHNOLOGY AND VIRTUALISATION

BY DEREK CASSIDY

Knowledge based economies are now becoming the corner stone of the recovery that we see emerging from the recession, the worst the world has seen in many decades. This economic recovery is being led by the development and innovation in technology, and the sciences that has opened up a new chapter in the European economic revival. Knowledge sharing and the increased ability to interact with the internet has opened up the world to the understanding that the intelligence that encircles and entwines the internet can be manipulated to the advantage of the user and so create a virtual world to interact, learn and develop. The internet, or the medium which all virtual interactions take place, is also the tangible product that many industries and technology houses use to develop and distribute their own products, whether these products are tangible or virtual like music/video or information that can be accessed and freely interrogated. Other industries are also using the internet to thrive and see it as the new highway, replacing the old world's view of product distribution and sales.

As the internet continues to grow and with the ever increasing confidence that people have placed in its ability to be a secure place to carry out financial transactions, purchase and receive products and services, has led to a growth in its use, with specific attention being drawn to business to business (B-2-B) and business to consumer (B-2-C) interactions. This increase in online retailing has helped produce a record breaking Black Friday sales phenomenon, which was first introduced by the online market retailers, that has now developed into a week-long sales interaction between the on-line retailer, using online financial transactions, and the consumer, who is willing to take part in this interaction

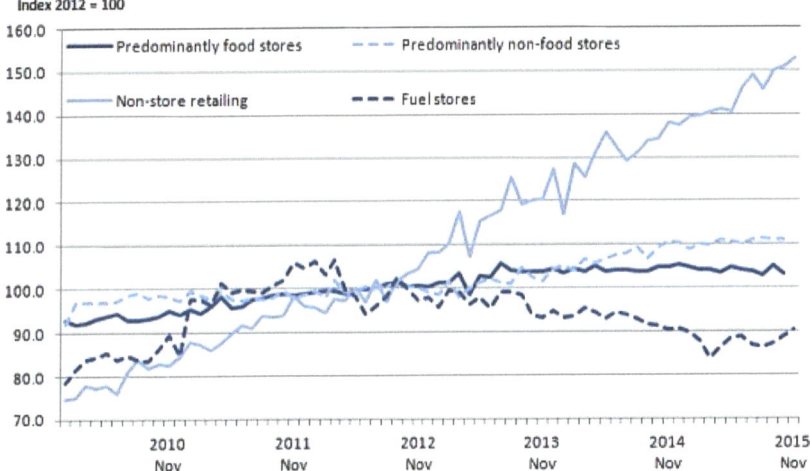

Graph showing increase in popularity of on-line sales over traditional sales: courtesy of www.thisismoney.co.uk

and bypass the one-time shop front retailer that we know today and develop new relationships with on-line virtual retailers that sell products and services that the consumer cannot touch and inspect but can only take on thrust.

However, it is not just an increase in internet traffic associated with commercial financial transactions, it is also the noticeable increase in communications like Skype and other products — two-way interaction with the likes of Facebook and Twitter and video manipulation of the internet like Netflix and YouTube, that has seen a sharp rise in capacity usage and bandwidth. However, the optical networks with their fixed infrastructure, being the transport backbone of the internet has not grown to meet the demand. This increased development in internet activity has its tangible draw backs and that is the slow pace of network design, development and infrastructure that is being built to carry this increase in traffic. The IEEE is leading the way as it completed the standardisation of the second generation 100Gb channel and is actively working on the first generation 400Gb channel so that the existing networks can increase capacity without the need to introduce physical infrastructure relying only on the upgrade of transmission

The growth of Ethernet speeds with regards to the IEEE: With courtesy of the Ethernet Alliance-www.ethernetalliance.org

and networking equipment to take on the capacity while keeping the existing optical infrastructure in place.

These new standards which have been agreed and standardised in the last few years allow for the network operators to incorporate 40GbE and 100GbE optical channels into their existing networks to increase the bandwidth available so that they, the network operators, could try and meet the ever

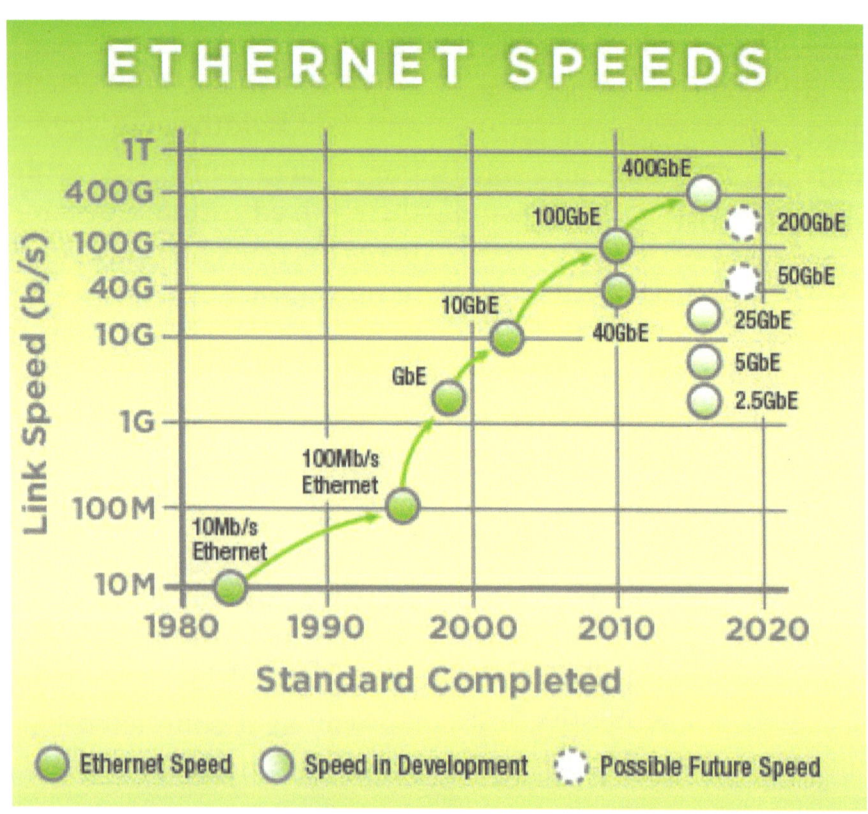

increasing demand required to carry the ever growing internet capacity. Existing networks have incorporated the first generation 100GbE optical channels operating at the original LR10 or 10 x 10GbE lanes, or the 4 x 10GbE lane multiplexed 40GbE. However, new network developments have allowed for the introduction of the second-generation 100GbE standard, which uses the multiplexed 4 x 25GbE optical lanes. But these developments do not stop here, the idea that 25GbE optical channels exist as multiplexing lanes for 100GbE, 200GbE and very soon 400GbE optical channels, has led to the investigation and standardisation of the 25GbE optical channel that has quickly opened up the idea that network operators to design new networks capable of delivering 25GbE single channels as direct client interconnecting links. These new developments into the single 25GbE optical channel has opened up the development and introduction of the 50GbE optical channels allowing for the network designers to look at ways of incorporating 400GbE optical channels into their optical network designs. These higher 25GbE and 50GbE optical Ethernet rates are in the near future but their introduction as a carrier at single channel or multiple channels as a 400GbE optical channel within a DWDM network is already forecasted.

Many optical carrier networks are now using first and second generation 40GbE and 100GbE optical channels as high order interconnects between networks and are looking at the development of incorporating these higher Ethernet bitrates into 100GbE DWDM networks and flexgrid DWDM -systems so that they can upgrade their next generation networks so that they can fulfil the capacity needs required to meet the increase in data usage. Some networks

have increased their optical channel capacity to 200GbE systems in the British Isles and have even gone as far as building 400GbE systems. They are now looking at building DWDM systems both foxed and flexgrid to deliver these channels as backbone carriers.

The Data Centres are also incorporating 100GbE multi-mode and single-mode optical channels at the core network level enabling for faster interconnect within the ISP and SAN storage space and also offering quicker access to core and backbone networks and the wide area networks (WAN) but there is a very keen interest to start using these as the standard interconnect as they reduce on increased infrastructure requirements to meet the same capacity levels. But with the continuous development and thrust in this technology it is foreseen that first generation 400GbE channels will soon make their mark among the Data Centre designers and also become the norm in optical channel usage within dense wave division multiplex (DWDM) networks, that are the back bone of our communication society.

With the global capacity in internet traffic to reach capacity levels that was never foreseen by the web designers or any other innovator the need to open up new pathways that can help deliver this capacity is underway. With the development of the new trans-Atlantic submarine cables connecting Canada to the UK with a link to Ireland and other independent cables connecting New York to Dublin to London will allow for a new infrastructure plans and designs to manipulate these new physicals connects to allow for the growth of internet capacity hungry industries and technological knowledge sharing societies to grow and meet the demand by being able to offer diverse paths to the markets enabling them to reduce the bottle necks and grow market share. With the future introduction of submarine cables from Ireland to France and the extension of existing submarine cables by independent operators to Europe from Ireland, this will allow for more diverse high-speed routes to enable the growth to continue.

Many countries, like Ireland and the UK, will benefit from these new systems that have incorporated into their designs the use of 100GbE technology so that they are capable of carrying multiple 100GbE DWDM systems per fibre pair allowing for a huge increase in channel capacity across the water and interconnecting Ireland with the financial centres of the UK, Europe and the US. These new systems will lead to an increased investment and aid the economic revival and recovery that is continuing today. The establishment of virtual data centres or cloud centres will allow major technological industry leaders like Apple, Google and Amazon to locate to areas where they do not need to be beside their industrial base, like to old industries which needed access to ports, rail and road. These new virtual data centres or cloud centres only need access to a high speed optical diverse networks that are reliable and can deliver the channel capacity that is required. It is only in the last few years, since the establishment of cloud services has it been possible for the large data centres to move away from areas of industrial concentration and build carbon neutral and green energy data centres that can be connected to and operated virtually enabling the growth of the cloud centre.

Also the growth in software-defined networks (SDN) is an aid to the realisation of the virtual and cloud storage networks that are being developed and growing in number. The ability to build and design a network that can be manipulated to deliver the required capacity or to reroute or define capacity when traffic is growing, which will allow for the networks to meet the demand and to develop future strategies that will enable the growth of the internet to me met head on with new transmission technologies that will create a true knowledge sharing society that has its framework on e-learning, e-business and the ability to interact. The internet, for so long was just a tool to be used as connect to data bases, but now it is the backbone of our communication systems and is at the heart of our society. The future demands of our optical carrier networks are to meet the every growing capacity requirement by making sure we have the optical network capacity in place to meet them.

Derek Cassidy is from Dublin, Ireland. He has worked for 23 years in the telecommunications industry of which 21 years have been spent dealing with optical terrestrial systems and submarine networks. He works for BT in their Networks and Field Engineering division. He is Founder and Chairman of the Irish Communications Research Group, a voluntary organisation dedicated to the promotion, protection and research of Ireland's communication heritage. He is a Chartered Engineer with the IET and Engineers Ireland, Chairman of the IET Ireland Network and is also a member of the IEEE, holding a position of Head of the IEEE (Ire) Consultants and Entrepreneurs Network. Derek holds the following Degrees; BSc (Physics/Optical Engineering), BEng (Structural/Mechanical Engineering) and BSc (Engineering Design) and has a Master's Degree MEng (Structural, Mechanical, and Forensic Engineering) and a Master's Degree MSc (Optical Engineering). He holds an HDip (Innovation and Design), Diploma (Electronic Engineering) and Diploma in Marketing and he is currently researching the Communication History of Ireland and is doing a PhD research programme in the field of Optical Engineering.

submarine telecoms FORUM

Sponsors Awards For
"Best Poster"
&
"Best Paper"
At

SubOptic 2016

www.suboptic.org

IS INDUSTRY CONSOLIDATION OVERDUE?
BY JAS DHOOPER

The telecommunications industry is no stranger to consolidation. Many investors, manufacturers and service providers will recall the late 1990's through to early 2000's and the massive restructuring that took place, particularly in US and European markets.

The routes to consolidation are invariably complex and this one clearly had too many players pitching for similar business, all trying to boost top-line growth whilst holding onto the same value-generating element – the customer.

A painful process of restructuring predictably followed suit and enabled those that survived in the sector to stay focused on core business competencies and scale back their operations to what was really needed to meet customer demand. In short, a new 'norm' was established in the necessary move 'back to basics'. Although less visible, the subsea telecom industry also took a battering.

TROUBLED WATERS AHEAD

In more recent times however, there has been an interesting new development — cash rich players from outside the sector, such as Facebook, Alphabet (Google) and others, are recognising the strategic value of owning telecommunication infrastructure, particularly the type that best facilitates international connectivity.

Their focus has remained simple — to move large amounts of data between their own data centres and simplify the value proposition. This could be seen as an early play towards building a MVNO, or even an Operator model — the value proposition being to better serve their increasingly global customer base at a cost, which is desired to be lower than that of even their traditional suppliers.

These relatively new organisations, with new money, ideas and ownership models, moving into non-traditional sector, are making subsea telecom an attractive new proposition. A change is in motion.

DATA DRIVEN DEMAND

Today, one fundamental aspect remains constant — the demand for regional and global bandwidth continues to grow at an expanding rate and it's a global phenomenon, cited from various sources including Cisco VNI

- Annual global IP traffic will surpass the zettabyte (1,000 exabytes) threshold in 2016. This has increased more than 5-fold during the past 5 years alone and expected to increase 3-fold over the next 5 years.

- Business IP traffic will grow at a CAGR of 18 percent from 2013 to 2018, driven by advancements in video communications.

- The fastest growth is expected in the Middle East and Africa, with a CAGR of 23 percent.

There is, at least so far, no evidence to suggest that any reduction in data oriented bandwidth demand is on the horizon. Operators are becoming increasingly reliant upon broadband internet services, triple and quadruple plans with the mobile element now forming a critical part of any major telecom growth strategy.

Vodafone, for example, is already moving in this direction, and offers so called 'quad play' in four European countries. TV content is also planned later this year as it moves closer to Virgin Media in an attempt to cut its reliance on rival BT.

CONTENT IS KING

Driven by advancements in broadband services, 4K TV, 4G mobile and not forgetting worldwide Cloud services, continue to fuel the capacity demand side. Whilst wireless internet continues to grow, so too has fixed line technology and internet access to the home.

Netflix is one prime example of how home entertainment has evolved in recent times. Whist their users consume large bandwidth, the application of CD caching technologies placing content closer to the end user, makes it more of a domestic play in most cases.

Coupled with mobility, the changing landscape of how we work, play and connect will have far reaching lifestyle and social implications in the very near future.

The next generation of 5G technologies will create another step change. For HD video, 5G will allow download an eight-gigabyte HD movie in *six seconds* versus the seven minutes it would take over 4G or more than an hour on 3G, according to a major industry supplier. Some interesting applications (and new revenue generating opportunities) are no doubt going to emerge.

As being promoted by Google, driver-less cars will need low latency (time it takes one device to send a packet of data to another device), currently for 4G, its around 50 milliseconds, but for 5G will reduce that to about one millisecond.

Video is often referred to as the main driver of internet bandwidth and in most regions more viewers are adopting mobile devices and tablets to consume such content. As reported in TSA, organisations such as Amazon are now challenging the traditional television industry and they are not alone.

The other major driver relates to the global demand for 'live' services – events like broadcast, new, sports event among others drives international bandwidth.

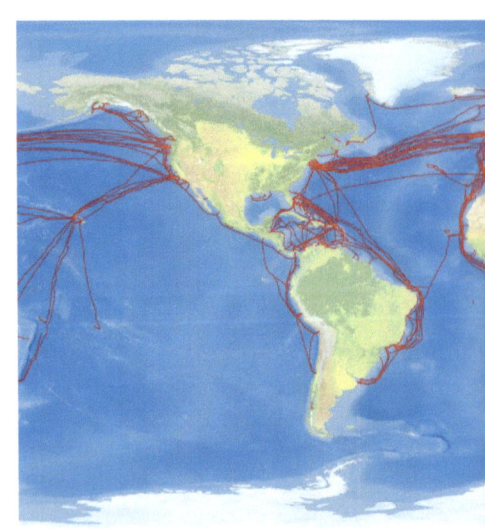

For the end user of course, it should be great news. They get even more choice that in turn drives creative packaging/bundling and price reductions, as each provider tries to dominate and hold onto their best customers, and create more of them. Some may say, however this is less choice as customers signed up to deals and are effectively locked-in. At a macro level this could yet lead to another memorable wave of consolidation and M&A in US and European markets. These mature markets allowed voracious competition that heavily resulted in market expansion, how many of them actually made a profit remains unclear, as many had to them sell assets to survive.

GLOBALISED NETWORK DESIGN

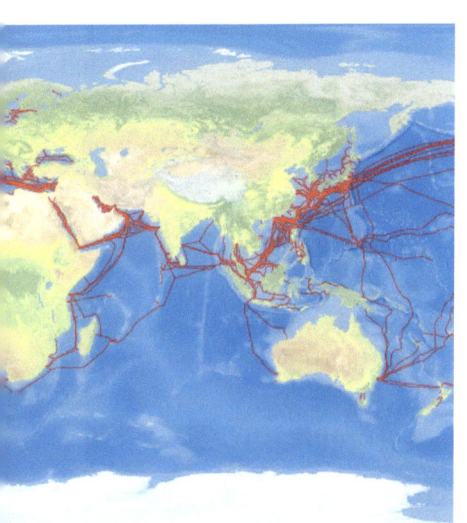

The merging of terrestrial telecommunication networks and undersea submarine cable systems has begun. As global demand for seamless connectivity increases, so does the need to mesh such networks together and service a growing need for high capacity, low latency services, offering robust end-to-end quality. High frequency trading has often been the driver for low latency networks, although other applications can also gain advantage.

So-called "high-frequency trades", controlled by computers, involve making what may be hundreds of thousands of transactions in less than a second — all determined by a program that tracks market conditions. Banks and hedge funds often use such applications to make rapid trade and enhance profit.

THE COMPLEX WEB – COURTESY OF GLOBAL MARINE SYSTEMS, UK

Beneath the world's oceans, seas and rivers is a network of underwater systems silently connecting even the remotest parts of the world to everyone else. They carry the world's internet traffic reliably and relentlessly every microsecond of every day, often only visible to

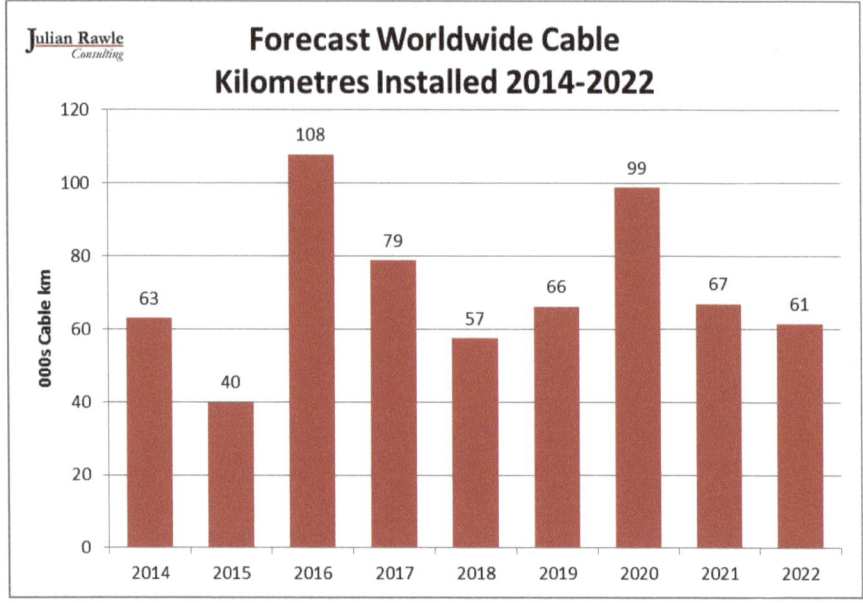

the public when there is a rare but catastrophic event, such as a cable cut, the effect of which isn't instantaneously healed by some of the most advanced redundancy and re-routing technology ever deployed.

Since 2012, the installed base of submarine cables has almost doubled from 150 to the current figure of 285, and 2014 has also been a huge year for the submarine cable industry, with new awards for the build out of new systems reaching a new level driven by significant investment.

ECONOMICS OF SUBSEA

Alongside these demand and supply conditions, the cost per km continues to come under increasing downward pressure, with the number of new projects increasing. Demand to service new geographical locations and deepen the World Wide Web to remote locations being the market pull.

With a backdrop of increasing telecom revenues, but squeezed margins and cost control within the service provider industry, some major players with strong balance sheets, are able and willing to fund consortium-led cable systems. There are several examples of large-scale systems (SMW-5, AAE-1, BBG, SEA-US, etc.) that have been deployed with this model during 2014- 2015.

GROWTH CONTINUES TO

MAKE THE MARKET ATTRACTIVE

It requires each stakeholder to contribute funding, in return for a percentage of capacity, with those with larger contributions, having greater ownership, through such investment models. TAT 9 cable system across the Atlantic had 36 members in its consortium during the late 1980, compared to Sea-Me-We-3's consortium numbered nearly 100.

It not only applies to international carriers, but also to private investors and service providers, like Vodafone, Digicel, and Hibernia, as just a few examples. Some can't sustain the growth, such as PacNet, acquired by Australian giant, Telstra, last year.

There is however much complexity within such models — a delay or show stopper issue with one stakeholder, or political unrest in one geographical location can add significant delay, perhaps even grind implementation to a halt. Thus, for turnkey suppliers, the project risk can be quite, if not too high, depending how the contract is structured

As such, this renewed interest in funding such submarine networks now represents a new wave of investment but also business uncertainty entering this market, since the industry shake up, that took place some 15 years ago.

As reported in the Wall Street Journal, Google Inc. is considering an investment in a new cable across the Pacific Ocean. Prior to this there was significant investment into Data Centres, thus laying the foundation to create a more private network. A fully connected world would allow the often-cited vision of the internet of things (IOT) to grow, develop and get another step closer to reality. In 2013, market value stood at $1.9 Trillion Projected to reach $7.1 Trillion by 2020 based upon market report in 2014.

A CONNECTED WORLD

There also seems to be a desire to reach out and provide broadband services to remote populations (preventing a digital divide of 'haves & have nots'), Africa being cited as one key example. Communities when connected, drive rapid economic growth, with some regions solely dependent upon tourism, broadband internet is a key enabler in this growth equation. Others have cited it as a means to help relieve poverty through such economic development, which over time builds in GDP growth.

Vital services such an education, e-learning and remote medical services can similarly accelerate on such a platform. The International Telecommunications Union (ITU) for example, has now called broadband internet access "a basic human right".

THE NATIONAL INTEREST

Whilst tapping into undersea fibre cable is pretty tough if not impossible, it has been cited in various press and media, that Governments have targeted cable landing stations to access content. These landing stations are termination points for ingress traffic so making access possible, although in principle, only with authorisation from the owner.

Private networks should offer improved security. Devices from outside the network cannot access these. Thus, "controlled" ownership provides value, which enables organisations to better serve their business model and growing demands. Management and security of content is critical to their value chain and in serving their customer base. Customers are also becoming increasingly aware of their rights, in term of data and privacy.

Many undersea cables were damaged by a powerful earthquake that hit Japan in March 2011

Cyber security concerns however are growing above and below sea level, as organisations and Governments invest in long-term plans to battle against such threats. It's been reported that a massive 90 percent of large business in UK have reported to have suffered an IT security breach. According to Forbes, this all comes at a cost to the business, as well as its shareholders (http://www.forbes.com/sites/dinamedland/2015/06/02/uk-government-urges-action-as-cost-of-cyber-security-breaches-doubles/).

Facebook, on the other hand as cited by various internet news feeds has invested an unspecified sum in the US $450 million regional subsea cable Asia Pacific Gateway (APG) — the purpose being to secure some guaranteed capacity on the 10,000-kilometer submarine network that will connect eight countries in South-East Asia. They too will need to address customers increasing content concerns, whist not owning the entire system.

It is apparent that such players intend to enter the subsea cable market to take advantage of strategic ownership, ahead of what some expect to be a desire to build and own end-to-end global connectivity. How will they manage security and government needs for access, remains to be seen, as even the largest players will struggle to ever own the entire system and all its connection points.

KEEPING LIGHTS ON

It is encouraging to see telecommunication infrastructure grow and aid global GDP growth. Whilst this critically supports the notion of a future "internet everywhere" as the end game, it also highlights our reliance on a reliably connected world.

For example, the Asia-America-Gateway (AAG) cable system has since 2009, experienced significant outages. Based upon various reports, these have been located at the intra-Asia segments of the system. Such dependence highlights the pain when a country experiences slow or no internet connectivity, as the application of satellite communication is often a costly alternative.

EXPOSURE TO THE UNCONTROLLABLE ENVIRONMENT

Threats to subsea cable systems range from accidents, political instability in key countries, and direct attacks on the cables. Such events can be become very problematic if quick resolutions are not found. The planning and design of such systems, recovery plans and even duplicated routes would seem a must have for nations with mission critical services, as repair operations can often take several days or more to complete.

Environmental factors (weather), permits (access) and vessel availability amongst other things can quickly render a repair plan redundant and add to the customer pain. Route engineering thus plays an increasing role in the design of such submarine cable systems at the outset of any project.

Duplication, disaster recovery strategies and operational readiness thus form critical planning activities amongst others. Higher end user expectations mean system owners now need too much more carefully consider and implement back office plans. The reliance of a cable vessel and its timely availability whilst covered by maintenance zones, such as ACMA (http://acmarepair.com/about/) offers some capability to manage the break-fix model and help restore service in acceptable timeframes.

To speed up repairs to such submarine cable systems, the maintenance fleet are normally on standby at strategically located ports, albeit as a shared pool of resources covering many cable systems.

ACMA was founded in 1965 and is a non-profit cooperative cable maintenance agreement, responsible for the operations and maintenance of undersea communications, as well as oil and gas platform operators, in the Atlantic, North Sea and South-Eastern Pacific Ocean.

In the past turnkey day rates for submarine cable installation assets could be up to $110k, whilst in today's environment they barely reach $85k. Significant landscape changes in the oil and gas market have brought about a drop in prices and demand. Market pressures from the oil and gas and telecom industries have also re-aligned to the demand curve.

MARINE INSTALLATION ASSETS NEED UP KEEP AND FEEDING

The challenge still remains however, regardless of which provider is used, of maintaining and ensuring that connectivity is consistent in today's demanding "always on" world.

SUSTAINABLE MARKET FORCES

Whilst the old days of boom and bust might not seem so likely to return, at least for now, there are other trends emerging which question if certain forces will lead to yet more industry consolidation.

The continual desire from purchasers to push down the per km per Gb cost in seeking better deals, could result in new structural changes within the industry supplier base. The idea to bundle in other services to bolster margins for supplying subsea cable systems is another strategy in a landscape of suppliers characterised by fierce rivalry and competition.

A mature industry characterised by cutthroat pricing is often an early sign for consolidation. With Buyers seeking aggressive pricing to meet their increasingly ambitious business cases, perhaps in the end they themselves drive this change.

Differentiation between players in terms of product seems to be less as most are building up or have a good track record. However, differentiation advantage depends on whether potential customers value the relationship with the supplier, and perhaps not just a price equation.

Cost control is always central to competitive strategy, whether an organisation is pursuing a strategy of cost advantage or differentiation. There are many solid examples of how industries have moved into lower cost manufacturing and labour to ride the cost curve.

CAPTURING THE VALUE

Differentiation, after all, is about understanding both the product and the customer. And the quest for advantage presents some fundamental questions for these companies: How do we create value for customers? And how do we do this more effectively and efficiently so as to earn a profit? What gives us the edge? How do we remain innovative in our solutions and agile enough to deal with an ever-changing landscape?

All of this needs careful tactical and strategic planning, working across varying time horizons and adapting to volatile environments as prediction becomes more difficult and the iterations between strategy formulation and implementation more frequent. The subsea telecom sector is evolving in a complex environment and inevitably subject to further change.

Product development and R&D is resource-hungry, but revenue growth depends on new products or new customers – often both are necessary. It is well known that innovation forms a key element of a supplier's differentiation in the market place. Standing still, with out dated products is not really a viable strategy any sector, unless of course the plan is to be acquired.

Subsea turnkey suppliers are now poised to battle it out to win new deals across the global inter-connected telecommunication market:

TURNKEY SUBMARINE CABLE SUPPLIERS – IS THERE ROOM FOR MORE?

Developing strategies to enhance top-line growth enhance margin and strike the right balance with profit and risk, simply gets tougher. An industry, some would say, perhaps not suited to the faint hearted.

Deals are won, at the expense of margin in the hope that they can recover some during the 25 year system life span, although the majority of customers would consider the use of alternative

NEC	Established market players with dominant regional positions
TE-Sub Com	
Alcatel Lucent	
Huawei Marine Networks	New market entrants
Xtera	

vendors, for upgrade capacity. The lowest cost route and best technology fit, being the primarily goal.

In addition during 2015, for example, Nokia have agreed to buy Alcatel-Lucent in an all share deal, which values the French rival at over $16 billion. The subsea element, known as Alcatel Submarine Networks, which intrinsically services global markets, is expected now to remain within the group. This could turn out to be a useful strategic move as it allows Nokia to reconsider with time options to spin out this separated business unit. Of course, having a parent can often be better particularly during tougher economic times.

Xtera a privately held organisation on the other hand, decided to go public as to raise much needed funds to grow and maintain it business plans, in 2015.

This is an interesting move that many are expected to watch, given it's the first within the subsea supplier base to take such a bold move. The stock market can be unforgiving when expectations are not met, as we have seen within the telecom sector. This move by Xtera, in time may reflect upon Nokia's forward-looking strategic plans with Alcatel Submarine Networks, story that is continued to develop.

Such deals do carry some uncertainty and risk and the track record of mergers in this sector has been poor. The process of actually gaining predicted value-added synergies on paper is always subject to the more practical implementation lessons from the past.

OPPORTUNITIES AND CHALLENGES

The subsea market in particular with each supplier bidding aggressively for a piece of the pie, is coming under increasing cost pressure that filters down the supply chain. Whilst controlling costs and moving operations into lower cost facilities is prudent in all highly competitive industries, the question about maintaining margin is a tough one – for a given risk appetite!

If there are not enough contracts awarded to each supplier to keep them going, then competition will become more intense for the available contracts and perhaps cause a shift to a buyers market?

Over the past 5 years, new entrants have emerged, challenging the dominant positions of the established few. Some degree of competition is typically good for most industries and warmly welcomed by most customers. Such entrants are bringing to this industry, new innovation and business models to tackle a historically attractive market. Some would say that this innovation and change has been lacking for too long.

Indeed, there are some suppliers that operate a fully vertical model and others that have adopted a horizontal one. Not that dissimilar to other industry models, in which costly assets (such as cable manufacturing and vessels) and manufacturing facilities are outsourced. As such, they don't need to be constant drain on the balance sheet and cash flow, nor need the organisation to maintain certain revenue to balance the books. Or worse, ask their shareholder(s) to dip into their pockets.

Complications however can arise with vertically integrated models, when top line growth suffers and the organisation is not able to adapt quickly enough to a changing landscape. Being asset light on the other hand can help in managing the ups and downs of the industry cycle, but requires closer operational integration and alignment. Some say, vertically integrated models offer better quality and control, after all the entire solution is 'in-house'.

There are indeed many trade-offs between such models, and having a mind-set and organisational culture to think long-term, navigate through a complex market place, and be willing adopt new models are all healthy ingredients for the future.

Above all what remains vital is cash to maintain keep the business going and cash to grow. Consolidations among the suppliers is an interesting complex equation and often in an overcrowded market, a more than likely path.

If the demand side keeps pushing to reduce the cost per km per Gb the ripple effect will result is fewer suppliers, perhaps through M&A or supplier collapse. Such industry dynamics were very apparent during the telecom boom period due to fierce rivalry and competition.

Can customers still get better services with a small set of suppliers? Will customers ultimately pay the price with increase supplier cost, when consolidation materialises? Will suppliers tend to retract to their home markets and perhaps, less dominate in other regions? Will

continues to provide upgrade capacity, as such multi-supplier systems are very much the norm. This present opportunities all round for additional margin that could pave the way to help to generate more business.

Either way, a complex situation is developing and ways to navigate through will be a key theme during 2016.

- Buyers of subsea cable systems remain in a strong position due to intense competition.

- Supply side continues to face increase rivalry among the supplier base, getting enough work to maintain the workforce is key.

Intense competition continues to grow as we progress into 2016 and beyond, it is clear each will have to continue to adjust and further change is a given within the supplier base. Each supplier has a business model and a cost base that needs to be serviced, in addition to making some kind of profit for its shareholders.

THE JOURNEY AHEAD

A healthy industry with many players and a capability to implement world-class subsea

some firm merger as to focus more on a sustainable model?

Whilst some suppliers can maintain lean operational models, and ride such a curve, it is indeed challenging to see how this can be sustained. In a market with less suppliers cost can often be found to increase.

How will the ripple effect impact cable and vessel providers being part of the supply chain? Will the cash rich players from outside the sector, such as Facebook, Alphabet (Google) drive the M&A equation?

No longer can it be assumed that the supplier that installs the system will be the one that telecom networks, manage turnkey risk and provide support will clearly give customers better choice.

In 2014 according to TSA report some 120,000 route kilometres were in contract position, representing a massive increase against the years before.

Positive for the overall market, but to avoid a recurrence of the 'booms and busts' of the past — there needs to be enough business for all suppliers! Committed and properly financed pipeline projects will be critical to survival and on-going cash flow needs.

Of course, survival is also about having a lean and efficient organisation, continually seeking efficiencies and innovation, among other key strategic initiatives and such are the challenges facing the industry today.

For now then, it would seem the need to service increasing demand in global capacity through subsea telecom infrastructure remains firm, driven by macro factors and new capacity buyers, making it viable and attractive for investment.

But convincing investors and shareholders, that the current landscape and contemporary business models are actually capable of generating sustained positive cash flow is of greater importance. More and many players might be good for choice, but less so for industry sustainability and stability.

It remains to be seen if the early stages of an industry consolidation in subsea telecom may have already commenced, but "watch this space" might be a prudent outlook as we sail into a new year.

Jas Dhooper has over 20 years' experience in the Submarine and Service Provider Telecom Sectors. Currently serving as, VP Service Delivery Office for Huawei Marine Networks (HMN) in China. He has gained significant management experience in setting up global capability to deliver large-scale multi-million dollar telecommunications complex programs, building multi-culture teams within ASIA, Middle East and European regions.

Jas also has held a number of senior management roles and technical positions in the operator side, working for Cable and Wireless (now Vodafone) and Interoute Communications since the mid-1990s and Global Marine Systems. During this period, Jas gained operational experience in Telecom M&A, Telecom network design and management.

Jas holds a Master of Business Administration (MBA), engineering honours degree from London University and has published several papers in the field of Telecommunications. He is a Chartered engineer, a Fellow member of the IET and member of the Institute of Directors, UK.

THE ATLANTIC: AN INFRASTRUCTURE ANALYSIS
BY HUBERT SOUISA

Global reach throughout the Internet is the main goal of network and infrastructure engineers. Although satellites are being used, the Internet depends mostly on undersea cables for data exchange between continents. These cables are of utmost importance to carriers, service providers, content providers and research networks throughout the world.

Whenever engineers begin to study the idea of building a new network between the United States and Europe, a primary concern is to ensure that the service will be robust and built on multiple underlying links so that if one goes down, users could still rely on sufficient bandwidth available.

In general, a network's transatlantic architecture will use capacity leased from the owners of three or four submarine cable systems (*Figure 1*). Unfortunately, any submarine cable faces a number of threats — from ship anchors and landslides, to undersea earthquakes.

Ship anchors cause most cable cuts near the coastline. Even though near-land undersea cables are designed with extra armor and are often buried up to a few meters deep, the anchors dropped by supertankers riding out storms can still damage the fiber optics inside the cables.

Figure 1: Northern Atlantic Submarine Cables

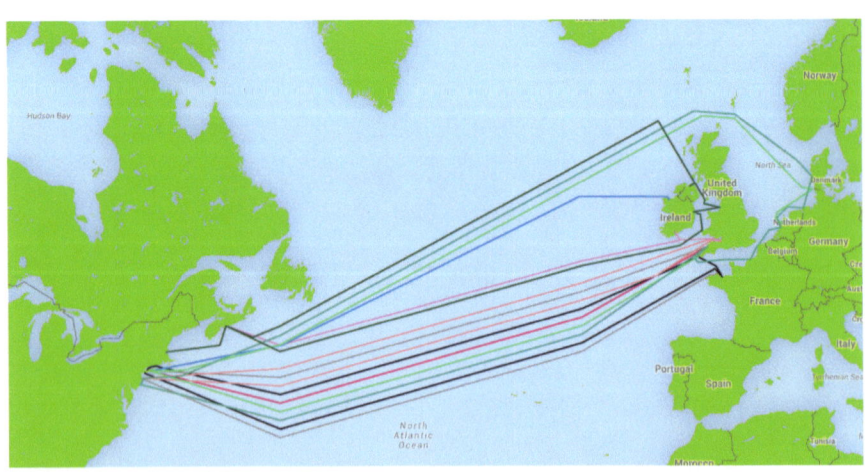

A major issue, then, is the time needed to repair a submarine cable break. On-land network owners can usually have an engineer at the site within four hours of a problem being reported and resolve it in a matter of days, or even just a few hours. With an undersea cable break, it can take a repair ship up to a week just to find the problem. It can take a few weeks to repair the cable, once located. The procedure can be delayed even further by storms and the limited number of repair ships.

The challenge for any network engineer is to analyze and select three or four linear Atlantic cable routes — each following a separate unique path or have the least amount of overlap — in order to minimize the probability of a complete network outage across multiple paths. Unfortunately, in your quest to find the safest routes for your network you might discover that there is no accurate database covering all cable paths, even though each operator knows the locations of his own cables. Despite TeleGeography's hard work on compiling the Submarine Cable Map website, none of the transatlantic submarine cable systems run in a straight line. However — based on the limited information available — the following transatlantic routing strategies are possible to provide three

Figure 2: Possible design until November 2015

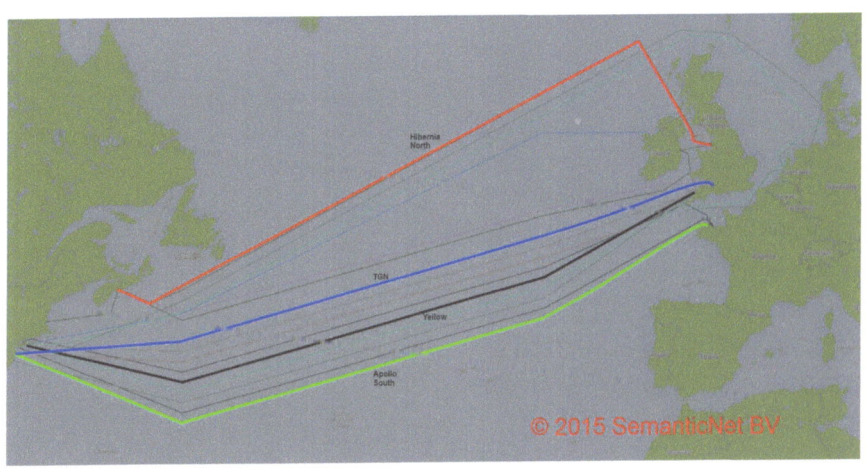

or four almost completely diverse paths.

PRE 2015

Up until the end of 2015, a robust network design would utilize two main paths between New York and the United Kingdom, while additional paths more north and south of the UK would supply far more security through physical diversity and different landing points (Figure 2). For example, one could incorporate the Hibernia Atlantic North cable between Canada and the UK and combine it with the Apollo South cable system. This would provide a two-route, geographically diverse network that was fully redundant between the United States and Europe. With a direct landing on the mainland of Europe, the Apollo South cable bypasses the London Hub, which is becoming a single-point-of-failure location for most international orientated internet organizations. Between New York and London, both Level 3's Yellow and Tata's TGN Atlantic cables could supply capacity and lower latency. This design offers protection against potential equipment and Cable Landing Station (CLS) failures anywhere in the region — especially in the UK. This is why it is important to utilize the Apollo and Hibernia systems for restoration purposes as well as a working path during normal operations.

Energy Sciences Network (ESnet) uses this design, as described by J. Metzer in a presentation held on October 15, 2014 [1]. This presentation describes ESnet's strategy to use 100G circuits on the Yellow, TGN, Apollo South and Hibernia North (40G) cable systems to support their extension to Europe's science networks. During the process of selecting the submarine cable systems you would like to use for your network, the choice can depend on the number of times cables cross each other on a certain route. Although well documented by the various cable operators, some engineers at organizations such as ESnet [2] may prefer lesser crossings on a cable, since multiple overlaps in a region could decrease their own network's resilience. One of the tools to help support a decision is the Kingfisher Information Service — Offshore Renewable & Cable Awareness project's online interactive map of the European region [3]. Analyzing the data reveals the number of possible overlaps and crossings between the transatlantic cable systems on the Eurasian Plate (*Table 1*).

POST 2015

As of November 2015, it is possible to use the new Hibernia Express and AEConnect cables (*Figure 3*). Both systems will enable the use of 100G and future wavelength technologies, which will be crucial in the further commoditization of the Atlantic capacity market. These systems offer an alternative for those looking to connect directly to Ireland or a diverse route through the north. These new cables have not changed the general network strategy for three or four paths across the Atlantic much, although some heavy users in the market might now have the need to bypass London and connect directly to the European mainland. However, current systems are getting older and at some point in time, capacity needs to be replenished — therefore, justifying new builds on this New York-London route.

On the northern path, the AEConnect system can be chosen as a replacement of the older Hibernia Atlantic North cable. With a carefully designed backhaul network, you could increase your

Table 1: Estimated Cable Crossings Eurasian Plate (number of crossing for AEConnect is unknown)

Crossings (estimated)	AC-1 North	AC-1 South	AEConnect	Apollo North	Apollo South	FA-1 North	FA-1 South	Hibernia Express	Hibernia North	Hibernia South	TAT-14 North	TAT-14 South	TGN-Atlantic North	TGN-Atlantic South	Yellow
Transatlantic Systems	2	1		3	1	5	3	2	2	5	2	3	4	5	4
Other systems (local/other regions)	10	4		4	6	4	7	3	9	4	13	4	3	4	4
Total Crossings	12	5	0	7	7	9	10	5	11	9	15	7	7	9	8

©2015 SemanticNet

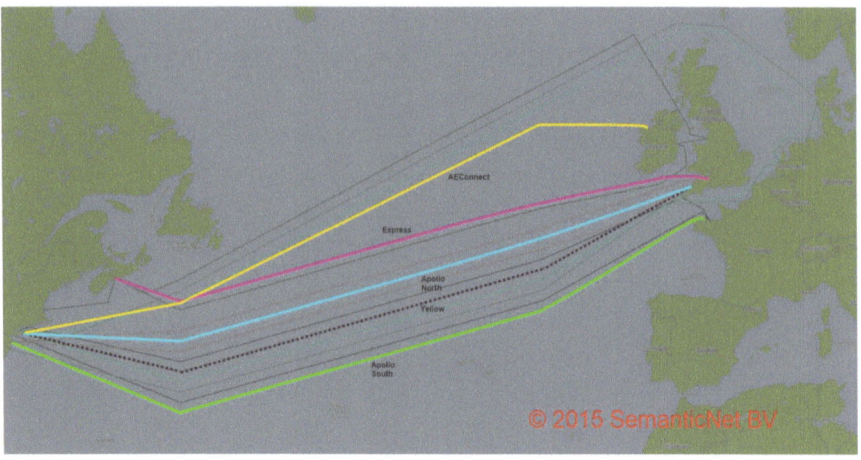

Figure 3: Post 2015 design

network's resiliency as well as directly connect to the Amsterdam Internet Hub and bypass the London metro. This may be critical because, for some network engineers, the Hibernia Express cable, combined with cables nearby such as Apollo North and/or Yellow, may not supply sufficient geographical diversity in the South West of England.

The main route between New York and London the Hibernia Express system — which lands close to TGN-Atlantic's cable station — might be better positioned to supply capacity for those who are looking for long-term contracts and future protection against technological advancements, such as 400G support. If you are looking for cheap, short-term contracts, it might be good to consider the TGN-Atlantic as well. For the third and fourth routes across the Atlantic the Apollo and Yellow cables are able to supply 100G circuits. Since AEConnect follows a less diverse path than Hibernia North either the Apollo South or FA-1 South cable is required in order to supply the much needed diversity through France and avoid any possible overlap with other cable systems and backhaul networks in Long Island (AEConnect) or the UK (Hibernia Express).

Alternatively, one might consider using only Hibernia's Express system and terminate at the Cork CLS in Ireland for increased

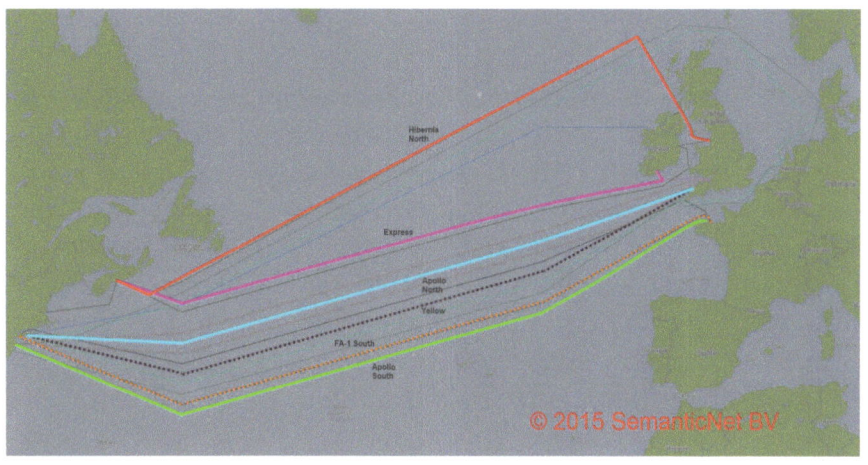

Figure 4: Alternative design

diversity from other cable landings in the UK (*Figure 4*) and combine this with the Hibernia North cable for redundancy in the Ireland/UK region. With a well-designed backhaul network from Cork to London and Amsterdam, the Express system is able to offer sufficient diversity in case of faults in the UK or New York region. For the third route, either Yellow or Apollo North could be selected — while the fourth route would again uses the Apollo South or FA-1 South for network diversity reasons and direct connections to the mainland of Europe.

Ashburn Hub & Future Diversity
When organizations need to connect to the increasing number of data centers in the Ashburn region, engineers will start to notice that all possible combinations of cable systems either land in New York or New Jersey. Those who are looking for connectivity to this Internet Hub might want to consider using the Apollo Cable System (*Figure 5*). This system offers termination points close to Ashburn as well as sufficient diversity via New York.

Alternatively, you could consider purchasing capacity on the TGN-Atlantic cable system or adding capacity on the AEConnect cable for more diversity to the European continent — although it shares a common backhaul on Long Island with the Apollo North cable.

Considering the growth of Internet hubs such as Ashburn, Miami and Paris, new systems offering direct access with onward connectivity to New York, Amsterdam, Frankfurt and London will become more important for closed user groups as well as service providers offering cloud and Internet services. The development of these systems will be driven by the need for more physical diversity between cables, landing stations and backhaul networks, as well as the ever-increasing need for capacity. Such a new cable route design is utilized by the Fibre Atlantic cable project [4], initiated by SemanticNet. This system is designed to increase the number of paths in the Atlantic, connects the Ashburn region by using a unique diverse route south of current cable systems, connects the continents via two new landing points in Virginia, USA and France. Another possibility might be the build of a direct cable between Miami and Western Europe, offering lower latency between the south of the US and mainland Europe.

CONCLUSION

In the past, your transatlantic network design would utilize three or four paths across the Atlantic. While two new cables have been built, this strategy has not changed much. However, some diversity is becoming

Figure 5: Possible Ashburn/Washington Strategy

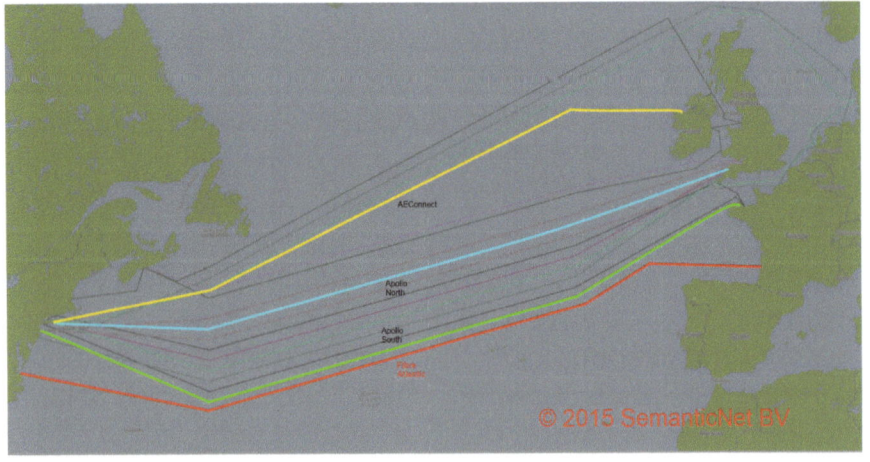

a more important driver for new cable projects, which are more south and physically diverse from the high-populated New York–London cable route. Ideally these new systems will change the resilience of the Atlantic route between Europe and the United States for the better. Potential paths include:

1. New York/Canada - North of England – mainland Europe
2. New York – Ireland/England – mainland Europe
3. Virginia – France
4. Miami – Spain/France

Combined with well designed backhaul networks that connects the key hubs on both continents, this would create new possibilities for OTTs, Carriers, Services Providers and NRENs to build meshed networks offering sufficient capacity, reliability and resiliency for the future.

Hubert Souisa is the founder and CEO of SemanticNet: a developer and provider of digital infrastructures. Supported by a strong team of industry specialists, SemanticNet is developing an alternative cable route in the Atlantic between the USA and mainland Europe, in order to increase overall connectivity, resiliency and availability and support Europe's Digital Agenda ambitions. For more information: www.semanticnet.net

submarine telecoms
FORUM

2016: A YEAR OF 'RADICAL' CHANGE IN SUBSEA SYSTEM DEVELOPMENT?

BY JOHN TIBBLES

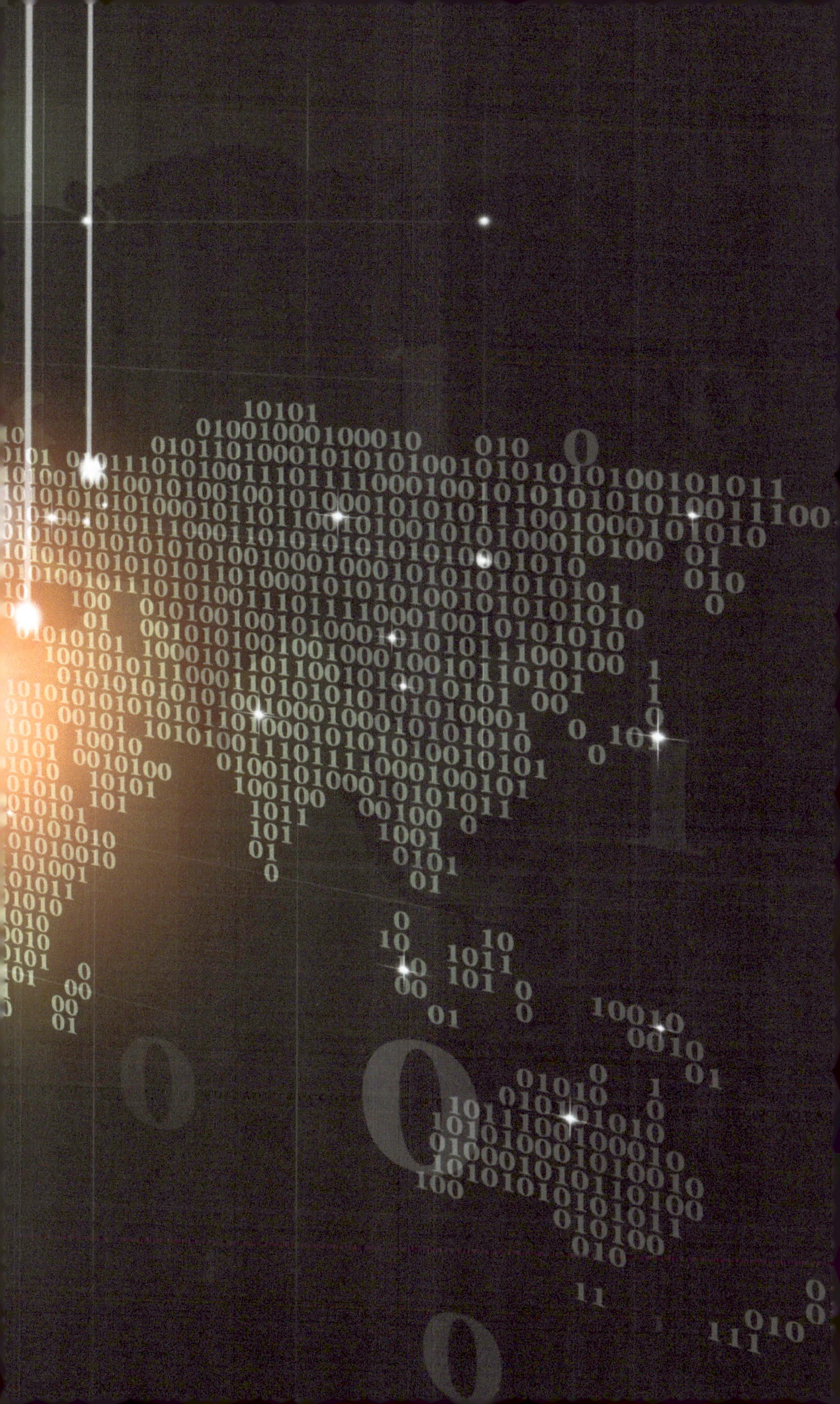

After thirty years involvement in the submarine cable world, it seems to me that we have reached something of a changing-of-the-guard situation with new entrants on the user and supplier sides of the equation. But, they haven't just entered the market. Within a few short years they are the key parties, on both the supply and customer side. That, to me, is a radical change so… What do I mean by radical?

In the submarine cable industry, we don't as a rule do radical. So I am not looking at a sudden and dramatic change, but a series of changes which have come into the industry in the past few years, which I think in 2016 will start to become the norm rather than exception. I am also focusing on the Western hemisphere since traditionally it is the Atlantic region that sets the pace for what happens in the cable world. That doesn't mean the Pacific Rim will see the same changes and, indeed, the continued appetite for investment by major carriers in that region is in direct

contrast to the behavior of the traditional Atlantic titans. It is a global industry, though, and ideas do spread out over time across the globe.

The submarine cable industry has, through necessity, always taken a cautious approach. Projects can be extremely expensive and by their nature cannot easily be modified or altered once in service. As a result, proven technology has always trumped cutting edge. Nothing wrong with that — it's a prudent and risk averse approach. Over time, however, this inherent caution leads to the development of a standardized one-size-fits-all approach and for many years that meant buying a system from a supplier of proven expertise, purpose built cable stations and one or more backhaul links to a network node often some distance away. It was proven and it worked and although it was tweaked for regulatory needs to allow more access at the cable station, it was pretty much the standard pattern used worldwide.

In recent years, though, we have seen significant changes in technology and network architecture on one side and a change in the nature of business and demands of users direct, like the content companies, and indirect, like the low latency hungry traders, on the other side. Changes have filtered into the industry piecemeal in recent years, but it looks to me like 2016 is the year when they all come together and directly and perhaps truly radically impact the technical design, network architecture and commercial construct of any new build system.

THE CUSTOMER FINALLY COMES FIRST

I am not being disparaging to the suppliers in saying that this has never really been the case. Traditionally, the suppliers have tried very hard to meet customer demands on timelines and capacity, but the truth is cables take a long time to manufacture and install and it's not wise to hurry either job. As to the product the basic issue was capacity and that was the art of the possible not the choice of the user. If you contracted to build a cable you got what technology, consistent with the critical issue of a very long warranty, could provide

The customers were almost exclusively carriers — only carriers could take the risk on investments that large because they were the only parties with requirements for large amounts of capacity. That was their business, they carried a lot of aggregated services such as voice, private line and corporate network data and in more recent years ISP traffic. Barely any end use entities could afford to invest or purchase directly even if regulations allowed.

There were two basic development models for cable projects: consortium or private. Both had their variations; — essentially one was a cooperative cost sharing enterprise among carriers and the other a private venture project with, or sometimes without, carrier participation taking advantage of external debt financing.

One was safe and one had some risks attached but either way the carriers were both purchasers and the end users and wanted the subsea capacity to integrate with their own — usually domestic — terrestrial networks. In other words either way the cables were built to serve the needs of the large carriers who were the biggest investors or customers depending on the nature of the project.

The first hint of a change to this scenario was the Unity project, a sort of limited invitation consortium, which included a non-carrier, Google, among its partners. A nice concept without the widespread voting complexities of traditional consortia but only really workable on extremely busy routes because the minimum

investment unit was a fibre pair — a substantial shift in itself away from the traditional widely inclusive consortium approach.

Private systems courted big carriers as anchor tenants and so also designed for their needs and in many cases had big carriers as part of their shareholder structure anyway.

Google of course were the precursor of what was to come-big data/ OTTs/content companies, whatever one chooses to call them; but parties like Google, Facebook, Microsoft Amazon and Equinix. They were different from what came before because their individual requirements were, and are, often greater than the aggregated needs of a large carrier and inevitably they asked themselves why buy cable capacity from someone with lesser requirements than me, why not buy it myself?

And the answer to that question was why indeed and these parties began to look to source large-scale capacity, typically a fibre pair on major transoceanic routes. In some cases this provided additional impetus for a proposed consortium type project and in others it was the catalyst that made the difference between launch and fail for a private project. And now it is rumored the next Atlantic system will go one step further and effectively be launched by at least two or three of this group of mega users.

Everyone knows the name of these entities, they are titans of Wall Street and modern business, so when they invest they drive a hard bargain when it comes to cost, but also when it comes to network architecture and system capability. They have big networks domestically for sure, but they are not configured like carrier networks they have their own established architecture linking their own datacenters and also with major 'public data centers' where they can interconnect with supplier networks or distribution networks to reach out to customers. And, of course these people are not stupid, they have seen the big carriers edge away from cable investments Just as they step in, they know they can make or break a project and thus demand costs and network architecture to suit them. Consequently private system developers have little choice but to go along because there is in the Atlantic region at least no one else to turn to for large-scale sales.

In addition, but on a smaller scale, there are a number of as yet untapped users who need to transfer large amounts of data around the world and do not want to pay what they perceive as high carrier charges just for capacity when they have their own internal technical competence and just need connectivity. These include supra national scientific and environmental projects and university/academic applications. These are not on the scale of the OTTs but they want the same kind of connectivity. In both cases they want the subsea capacity to be accessed not at a cable station, not at a carrier owned or even carrier neutral traditional PoP but at a purpose built and designed data center such as the ones run by Equinix and many others.

The OTTs at this point in time are essentially American, they have big domestic networks based on their own or leased dark fibre with their choice of optical transmission and NM technology at each end. Of course when the one of these links is a submarine one, that was not the way it has always been configured in the past when it was under carrier control. In the traditional cable world, the repeaters, the terminals were all supplied under an integrated contract. But the new projects where the customers buy a fibre pair,

they can and will put their own favored brand of optical transmission equipment and network management equipment on the end of the subsea link giving them a seamless coherent path from one end of their network to another without going anywhere near a carrier. Each path on their network is the same whether the nameplate over the data center symbol says Denver or Frankfurt it is just another link.

So as we are about to step over the invisible step between the first and second halves of this decade, for the first time the subsea cable industry has had to focus on the needs of end users. These parties now have the money, the plans and the power to get what they want and they know it.

NEW CHALLENGES FOR THE TRADITIONAL PRIME CONTRACTORS

So what does this change mean for the people who build the cables, the system suppliers?

Never a crowded market and now with only four really significant companies, it is a long time since contracts were largely awarded on the basis of nationality and politics. In spite of years of globalization, the four suppliers pattern of business does reflect some carry overs from the past in terms of regions where one is more successful than others, or regions where one is somewhat persona non grata and others where one is seen as more receptive to private system ideas.

In spite of these market skewing factors though, they all have to face up to the fact that change is upon them and they can no longer be sure that they will act as primary supplier and project integrator, assembling the cable itself, subsea optics, the terminal station equipment, power and SLTE as well as providing marine installation and repair services.

The reason for this change started in the recent past and the explosion of DWDM capability brought about by the use of terrestrial network based optics from terrestrial network suppliers onto the subsea systems. The economics of these super-upgrades are hugely attractive to system owners but in many cases they have not turned to the original supplier but to terrestrial networking specialists such as Ciena, Cisco and Infonera who had no tradition in the subsea industry but were familiar to the carrier community because of domestic network use.

These upgrades worked and proved that it was possible and practical to replace original SLTE equipment with one chosen by the consortium or system owner and in a few cases the owner of a fibre pair. It was a logical step to deploy equipment well known to them over the subsea cable paths. And technology available from these suppliers meant you didn't now need to buy a whole fibre pair but could use it on spectrum, itself another change to the established order. Spectrum allows a user to buy a virtual fibre pair, owning part of the spectrum it supports without the high cost of the physical pair itself.

This scenario has developed further with proposals to establish an ITU standard for open or 'naked cables' — that's to say, there is no proprietary SLTE. It is an idea embraced more willingly by some system suppliers than others but which seems inevitable that it will spread. It may even become the norm even though it will reduce the system suppliers contribution to the value chain and therefore their margins and income potential from new systems.

Even before the OTTs entered the market there were changes to the traditional way things were done which reduced the complexity and revenue opportunity of the supplier. Cable stations with their SLTE and interconnect capabilities are being reduced to small often unmanned power supply 'huts' with much of the traditional function transferred back to more conveniently located city PoP via landline fibre — as a concept not very different to terminating in a data center.

Of the four main suppliers, two are Asian based and manufactur a wide range of optical networking equipment — but at the present time these two are not among the market leaders share the bulk of the intercontinental and long haul market. Of these, one is TES/ Tyco and the other Alcatel Lucent Submarine Networks, now in the process of becoming part of Nokia Networks.

They both face challenges posed by the powerful new entrants to the customer side of the equation and the demands and expectations they bring with them to use their choice of technology. Of the two main suppliers: TES are not part of large conglomerate Telecoms technology business and will the pressure on them force them into an even smaller niche if 'open' cables become the norm in specifications for new systems

Alcatel are part of a much wider group with all kinds of telecom products but who have recently been on

the market without much if any serious interest. Not an ideal entrée to life in a new parent organization like Nokia, who have no experience whatsoever in subsea technology, but because of the failure of the efforts to IPO the cable division are likely to look for a new executive team to go forward with.

Both of these parties have found it a challenge to win upgrade business against the new entrants, they have succeeded in some cases but this market used to be assured revenue for them.

And so just at a time when significant change is facing the industry the small supplier group faces challenges of its own in keeping the subsea cable supply and integration viable and financially attractive. Although the new system market is quite buoyant TES and ASN have these issues to deal. In contrast the terrestrial optical networking specialist are determined to push forward.

From the inroads they have made with upgrades they are moving to be the prime optical transmission provider right from the start of a new generation of 'open 'systems, the subsea equivalent of dark fibre.

What does this mean for the suppliers? Are their parent companies going to be happy with lower margins and reduced market presence, are national or multinational political considerations going to impact freedom of the owners to sell or restructure what are in the end manufacturers of a product with serious national security implications? If critical elements of the subsea system are now made by a third party will they still be able to afford, or have the will to offer the assurance of lifetime system warranties, which are essentially what the customer has paid for in the past.

NEW TECHNOLOGY LEADS TO NEW PRODUCTS

For the cable developer technology advances provide new opportunities whether one is an OTT building a cable for the first time or a system developer acting almost as a proxy for the big users launching a private project. The emergence of spectrum management products allow virtual fibre pairs to be sold, something that seems to me vital in extending this open concept to routes, ultra-long haul or serving the developing world, where a whole fibre pair is too much even for the deepest of pockets.

The developer can also use spectrum products to optimize the fibre count he no longer needs to provide and try and sell a whole fibre pair to a big anchor tenant users-he can sell half the spectrum on the pair to one party and half on another. Alternatively he can retain unused capacity on part sold pairs and use it to deploy more conventional carrier targeted services, IRU sales and leases, or try and win the business of global scientific or research projects directly.

The developer can look to deploy a wider range of services based on advanced terrestrial transmission optic and related network management equipment and so spectrum products can benefit the customer and system developer alike. In fact because it makes the advantages of fibre pair ownership available at much lower demand levels it is a critically important new product that really can be a game changer in the way capacity is purchased. No more endless committee meetings and compensation for non-upgrading parties in a consortium, no more having to renegotiate terms from a private developer.

In addition newer optical transmission technology no longer needs the more complex compensated fibre necessary for use in a whole generation of DWDM systems. The ability to use a larger effective optical aperture fibre and other advances in optics technology holds out the promise of yet more bandwidth per pair than available from even the most recent upgrade projects.

So finally after years of end users saying what I really need is an end to end coherent global network where domestic and international facilities are integrated and operate to common standards via a single Network Management systems they are finally going to get it.

TO SUMMARIZE

For Traditional Suppliers

Will the choice of system supplier be less critical than before as traditional suppliers are forced to provide open access interfaces to their systems, providing only the cable and subsea plant, the power and simple SLTEs for testing only? Much of the critical innovative technology will be supplied by the optical networking segment that have the expertise and capability to provide and manage very high-level data rates and complex network management systems provide.

This on top of a very difficult period for the suppliers since it will commodities their product and lower margins in an already risky and competitive segment. Indeed with Alcatel unable to shed its ASN division before the Nokia take over how will the new owners see a potentially reduced value subsea cable product line. Similarly Tyco Corporation are not known for keeping hold of declining value product lines so they have their own challenges.

supply process may become equal or even above that of the subsea cable supplier. However, they have to deal with the warranty expectations of customers who hitherto have liked the prime contractor approach and a single provider of long extensive system warranties.

For Carriers

Already usurped by OTT/BigData/Contentcompanies as the big investors in subsea systems, what role is there for the traditional carriers in a world where Data Centers are the new 'Toll Switching Centre/ PoP. These new interconnect and 'switching points' no longer belong to carriers and nor does the international and often backhaul networks. Not only have they lost control of what was once their preserve they have lost it to potentially their largest customers –a double whammy. Long haul networks and sub cables may not have been in vogue in the big carrier domain of late with focus on FTTH and content for TV offerings,

Countering the possible purely corporate view of these divisions is the recognition that this is a critical state sensitive technology and possible governments on both sides of the Atlantic will seek to preserve these capabilities on that basis.

For the Optical Networking Experts.

Things can only get better as they are seen more and more as integral to providing a customer focused coherent system. Their place in the but is that all they aspire to? Some are looking to sell off their global networks and lease back what they require in a similar way to Cell tower sales but I suspect the complexities and multinational nature of cable systems will make that a lot harder

For global corporate users

The new technology developments and structural changes mean big corporations and multi-national institutions no longer need deal with the carrier giants. Just plug their LANs into the nearest data center to their main or regional offices and their LAN goes around the world as easily as round the office. (Well almost!)

For the Content/OTT sector

The new kids on the block will be the main beneficiaries of these trends and, as they have the money and the will to use their market power, more and more systems will be designed from the start to suit their needs. Will they enter the carrier market and trade, buy and sell excess capacity /bandwidth or will they shy away from that for regulatory reasons and allegations that they have become too vertically integrated. Indeed with the current members of this sector being almost entirely US based what view might the EU take of Trans-Atlantic communications being almost wholly US owned.

IN CONCLUSION

In itself, 2016 does not represent a radical departure from the past but it is the year in which a number of dtrends, which have already emerged, will come together to reshape the subsea cable market, indeed it is probably that further changes will occur and underline that trend.

Whatever happens during the year I think it will be seen as a defining period and on the conservative time lines our industry is used to it will indeed be seen as the year of radical change.

John Tibbles has spent over 30 years managing globally based investments in cable systems for some of the worlds major subsea network operators and owners involving strategic planning, partnerships and consortia management , buying and selling in the wholesale space and managing supplier relationships. He has been actively involved as a panelist, presenter and member of many industry bodies including SubOptic, PTC, ICPC as well as contributing to media articles on the industry. Now retired from daily involvement he owns JTIC consulting (www.consultjtic.com) providing consulting services for the submarine cable sector and the broader international carrier business

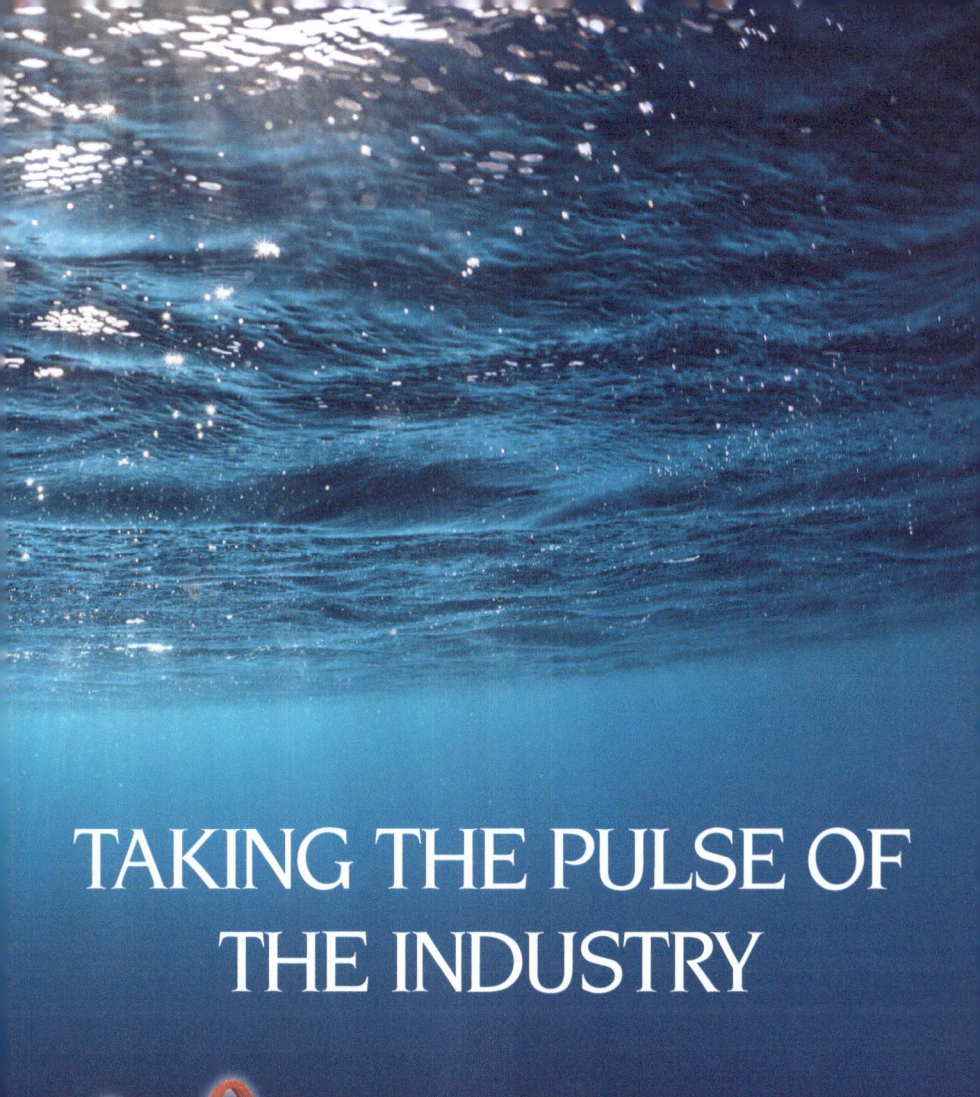

REALIZING THE VALUE OF A CONNECTED AFRICA

BY MIKE LAST

The transformational nature of the connectivity landscape in Africa has implications beyond the continent itself to the rest of the world.

Demand for reliable, high-speed connectivity throughout the world's second most populous continent continues to grow rapidly. Africa's 1.1 billion inhabitants and businesses increasingly take advantage of the many opportunities the global internet offers — in exactly the same way as in other internet-enabled parts of the world.

ATTRACTIVE ECONOMICS

The availability of reliable, affordable, high-capacity international connectivity is an important enabler to economic growth. It is no coincidence that after rapid expansion of the continent's submarine and terrestrial connectivity landscapes, nine of the world's 20 fastest-growing economies (by compounded annual growth rate, CAGR) for 2014-17 are predicted to be in Africa.

In 2015, Sub-Saharan Africa's GDP is expected to have grown at 4.5 percent, making it the fastest-growing economic zone in the world — outpacing Asia's regional average of 4.3 percent annual growth. Elsewhere, the GSMA reported that mobile telecommunications currently contributes over 6 percent of Sub-Saharan Africa's GDP and forecasts this will rise to over 8 percent by 2020.

In 2012, the mobile economy directly supported 3.3 million jobs and contributed US $21 billion to public funding in the region. By 2020, mobile is set to double its economic effect — employing 6.6 million people and contributing US $42 billion to public funding, including licence fees. The extent and longevity of this growth are evident in forecasts that Africa's mobile phone market will almost quadruple

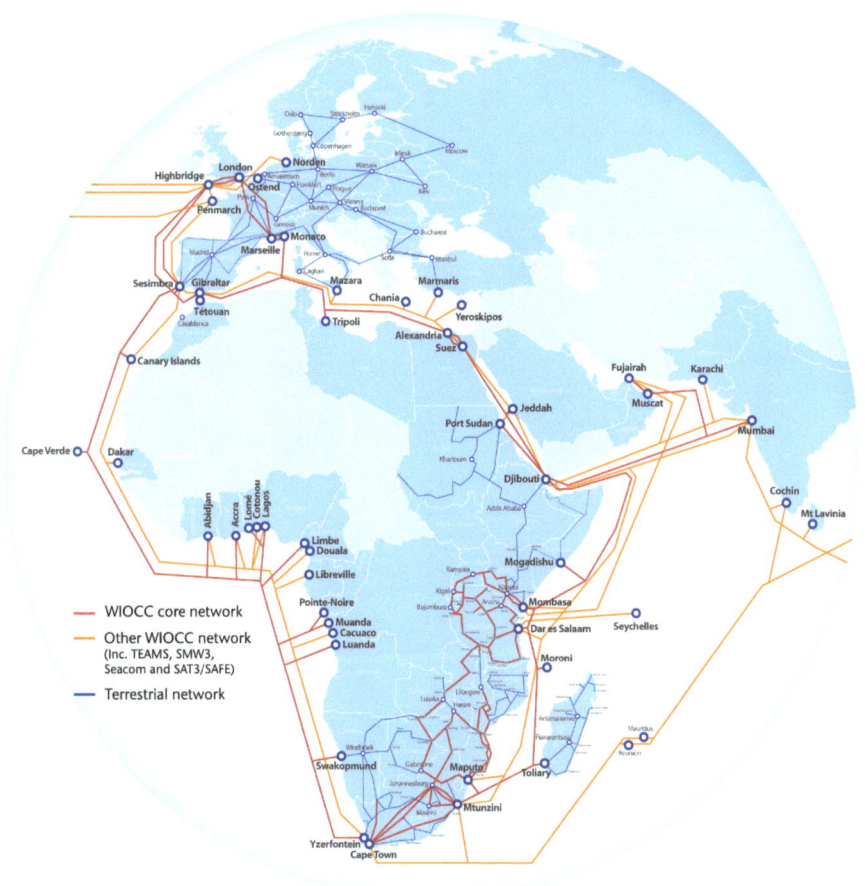

in value to $234 billion between 2013 and 2020.

Since early 2009, the number of international submarine cables serving Sub-Saharan Africa has increased from two to 11, and the total design capacity now exceeds a plentiful 60Tbps.

This improved submarine connectivity and rising user demand catalysed a surge in the growth and enhancement of terrestrial networks, making this international connectivity accessible to an increasing percentage of the population.

According to Hamilton Research, Africa's total terrestrial transmission network inventory has more than doubled in the last five years. The total

inventory of operational, under construction, planned and proposed terrestrial fibre-optic and microwave networks in Africa in June 2015 was 1 million, 19,643km.

Further investment will continue to improve connectivity significantly — bringing 53.7 percent (436 million) of Sub-Saharan Africa's population within 25km of an operational fibre-optic network node when all terrestrial fibre-optic and microwave networks — currently under construction, planned and proposed — enter service.

from all regions of the world. WIOCC's unique, diversity-rich pan-African network delivers direct connectivity to more than 500 locations in 30 African countries and guarantees sufficient scale and flexibility to support the rapid upgrades now required to meet ever-growing customer demands. International capacity wholesaler WIOCC continues to reinvest to further improve the quality and reliability of its network and infrastructure. In the past year, WIOCC has installed multiple Points of Presence (PoPs) in various African countries to provide customers with easy local access to the global internet.

55,000KM TERRESTRIAL NETWORK

The largest contiguous terrestrial network in Africa is 55,000km in length and operated by Africa's carriers' carrier, WIOCC, which has multiple customers — including many African and global carriers and ISPs —

Individual end-users are utilising this international connectivity for a host of purposes, including to access information; to communicate with others on Facebook, Skype, Twitter, YouTube, etc; to stream music and films; to handle and transfer money; for online gaming; in basic education; and to access assorted e-health and e-government services. Meanwhile businesses are using the improved connectivity to help them boost their efficiency and to open up new markets.

Improved connectivity is also opening up African higher education and research, with organisations such as the UbuntuNet Alliance working with WIOCC to link institutions throughout Africa. A lot of research is now being done in Africa, generating big data files that are of interest elsewhere in the world.

FURTHER BANDWIDTH GROWTH IN AFRICA

Internet penetration in Africa is only 28.6 percent — compared to 49.7 percent for the rest of the world — indicating that there is still significant scope for further expansion even though internet usage in Africa has already increased by a staggering 7,356 percent since 2000 (from 4.5million in 2000 to 331million in November 2015).

As more and more individuals and businesses in Africa have gained access to the internet, their reliance upon the wealth of benefits and services this provides has grown too. Diversity is, therefore, an increasingly important requirement, as carriers and ISPs strive to deliver their customers the levels of reliability they demand.

To exploit the significant commercial opportunities that now exist, operators and ISPs need to offer product packages that meet the reliability and increasingly bandwidth-hungry appetites of their customers. As a result, many are looking to increase the bandwidth, diversity and quality within their service portfolios, taking advantage of the cost-efficiencies, improved reach and greater reliability now possible through investments in submarine cables and terrestrial fibre networks.

UNLOCKING 'OPPORTUNITY AFRICA'

Local expertise, market knowledge, local contacts and an understanding of the different regulatory environments are all vital in order to be able to seize the opportunity within this diverse, 54-country continent. So too is the ability to establish long-term, mutually-beneficial partnerships, where a shared desire and ability to work flexibly together enables rapidly changing end-user demands to be quickly and easily met.

> **SINCE 2000, THE NUMBER OF INTERNET USERS IN AFRICA HAS GROWN BY 7,356 PERCENT**
>
> **INTERNET PENETRATION IN AFRICA IS 28.6 PERCENT — COMPARED TO 49.7 PERCENT FOR THE REST OF THE WORLD**
>
> **NINE OF THE WORLD'S 20 FASTEST-GROWING ECONOMIES FOR THE PERIOD 2014-17 ARE PREDICTED TO BE IN AFRICA**

By partnering with an experienced, Africa-based capacity wholesaler such as WIOCC, international telcos, carriers and ISPs can address their complex African connectivity requirements with a single organisation that is responsible for negotiating all the agreements and maintaining relationships with multiple operators in different countries. WIOCC's customer service, technical and network management experts also deliver seamless end-to-end services, which are pro-actively managed 24x7x365 from their Network Operations Centre in Kenya.

In 2014, WIOCC's partner-focused approach opened up Somalia — one of the last "corners" of Africa without direct fibre-optic connectivity — as well as delivering affordable, high-capacity international connectivity to landlocked countries such as Botswana, Burundi, Lesotho, Malawi, Zambia and Zimbabwe.

The resulting improvements in information access are helping transform life in Somalia, encouraging renewed investment from the diaspora and the international business community, which is creating a platform for sustainable economic growth.

Capacity take-up in Mogadishu continues to exceed expectations, and the flexible partnership approach is helping drive further cost reductions and catalyse improvements in terrestrial infrastructure.

PARTNERSHIP APPROACH

Partnership is without doubt one of the keys to success in Africa's rapidly changing telecommunications markets, and could differentiate the winners from the losers. Improved international connectivity, enhanced ICT infrastructure, more affordable high-performance handsets and prodigious growth in

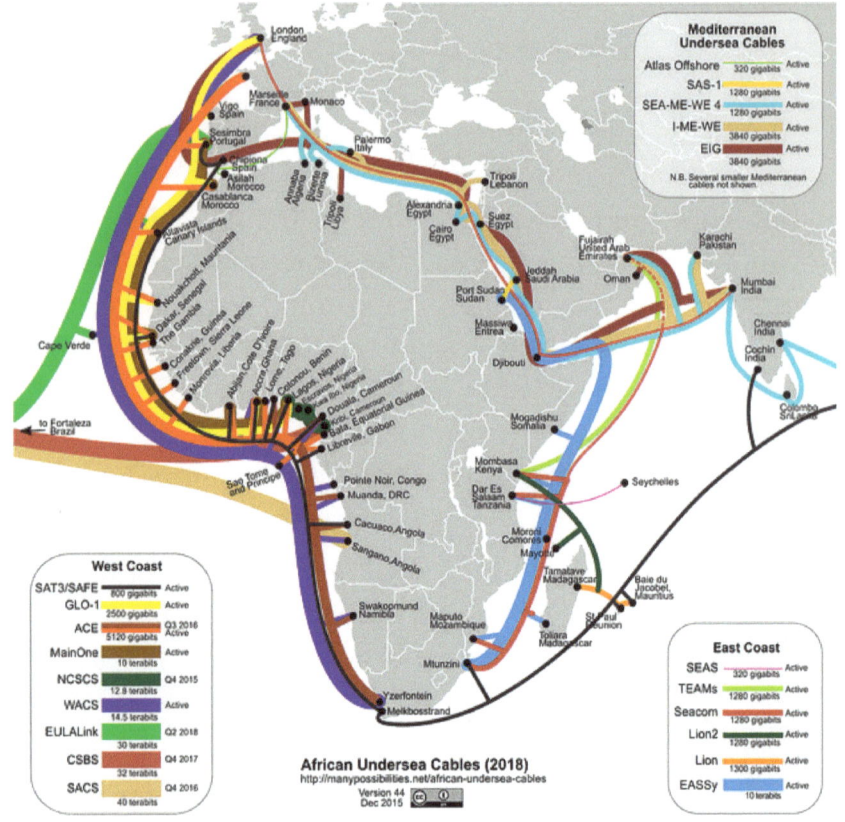

mobile, internet and data services are all having an effect — transforming how businesses in Africa operate and changing the way many individuals go about their daily lives.

These advances have helped improve business efficiency and productivity, opened up previously inaccessible markets, made eCommerce practical across more of the continent and led to the formation of many new companies and even business sectors.

The growing economic importance of Africa, combined with dramatic growth in demand for high-capacity bandwidth within the world's second largest continent, are encouraging an increasing number of international carriers, ISPs and telcos to expand their presence in Africa.

Historically, limited submarine and terrestrial connectivity restricted business opportunities for carriers and ISPs operating in Africa. However, over the past few years Africa's connectivity landscape has changed dramatically for the better.

IT'S NOT TOO LATE TO BE PART OF THE AFRICAN SUCCESS STORY

Delivering reliable, high-speed, international connectivity to, from and within Africa presents a rich commercial opportunity for both new entrants and existing players. There is significant expansion still to come before Africa reaches the penetration and usage levels of more established markets. The key to success is working with the right local partner now, so that your business is also a part of the African success story.

Mike Last is the Chief Marketing Officer at Africa's carriers' carrier WIOCC.

submarine telecoms
INDUSTRY REPORT

ISSUE 4 | NOVEMBER 2015

AVAILABLE NOW
http://amzn.to/1PKiK24

submarine telecoms
FORUM

Telecoms consulting of submarine cable systems for regional and trans-oceanic applications

ADVERTISER'S CORNER
BY KRISTIAN NIELSEN

A Year in Review

Dear readers,

Before we celebrate the hotly anticipated rise of 2016, please allow me to share a few milestones we barely registered during our mad dash into the new year –

In 2015, SubTel Forum was visited by its devoted readers over 10 million times and an average of about 800,000 times per month. That's roughly 20% more than 2014!

By the numbers, every single SubTel publication saw an increase last year – our single most downloaded product was the May edition of the Submarine Cable Almanac, downloaded 572,000 times. We introduced the Industry Supplement this year, which saw a limited run of four issues targeted at highlighting elements of the industry for laymen and industry outsiders.

Last year gave us one of the biggest changes to our publication line-ups – a partnership with Telegeography that enhanced our Industry Report in ways that, frankly, we had never imagined. When we originally created the Industry Report, we sought to create a light document that could give an overview of the industry. Four years later, the document is one of our core publications and is now supported by a dedicated team of analysts in a new project that we're calling STF Analytics.

While the crystal ball is always cloudy, I can say with

absolute certainty that we will maintain the level of analysis, opinion, and reporting that you, our readers, have come to expect. In 2016, we will grow the scope of that analysis with STF Analytics and our news room. Keep a keen eye on us, 2016 is a big, big year!

Ever loyally yours,

Kristian Nielsen literally grew up in the business since his first 'romp' on a BTM cableship in Southampton at age 5. He has been with Submarine Telecoms Forum for a little over 6 years; he is the originator of many products, such as the Submarine Cable Map, STF Today Live Video Stream, and the STF Cable Database. In 2013, Kristian was appointed Vice President and is now responsible for the vision, sales, and over-all direction and sales of SubTel Forum.

 +1 703.444.0845

 knielsen@subtelforum.com

Submarine Telecoms Forum, Inc.
21495 Ridgetop Circle, Suite 201
Sterling, Virginia 20166, USA
ISSN No. 1948-3031

PUBLISHER:
Wayne Nielsen
VICE PRESIDENT:
Kristian Nielsen
MANAGING EDITOR:
Kevin G. Summers

CONTRIBUTING AUTHORS:
Derek Cassidy, Kieran Clark, Jas Dhooper, Mike Last, Hubert Souisa, John Tibbles

Contributions are welcomed. Please forward to the Managing Editor at editor@subtelforum.com.

Submarine Telecoms Forum magazine is published bimonthly by Submarine Telecoms Forum, Inc., and is an independent commercial publication, serving as a freely accessible forum for professionals in industries connected with submarine optical fiber technologies and techniques. Submarine Telecoms Forum may not be reproduced or transmitted in any form, in whole or in part, without the permission of the publishers.

Liability: while every care is taken in preparation of this publication, the publishers cannot be held responsible for the accuracy of the information herein, or any errors which may occur in advertising or editorial content, or any consequence arising from any errors or omissions, and the editor reserves the right to edit any advertising or editorial material submitted for publication.

Copyright © 2016 Submarine Telecoms Forum, Inc.

submarine telecoms FORUM

January:
Global Outlook

March:
Finance & Legal

May:
Subsea Capacity

July:
Regional Systems

September:
Offshore Energy

November:
System Upgrades

Conferences

PTC 2016
17-20 January 2016
Honolulu, Hawaii USA
Website

ICPC Plenary Meeting
12-14 April 2016
Hamburg, Germany
Website

SubOptic 2016
18-21 April 2016
Dubai, UAE
Website

 SUBSCRIBE TO OUR FEED

 JOIN OUR MAILING LIST

Voice of the Industry

CODA
BY KEVIN G. SUMMERS

It's freezing cold this week in Amissville. The water troughs are frozen when I come out in the morning and I have to drain the hoses at night. Sure, we were wearing t-shirts and shorts last week, but that's another story. It's freezing cold this week, and that must mean that all of my co-workers are headed to Oahu.

Yes, PTC is here again and I've spent the past few weeks updating brochures, updating websites, and finalizing this issue of SubTel Forum magazine. And I did it all from my farm in the Virginia Piedmont, in the shadow of the Blue Ridge Moutains when the sun goes down. I spent a grand total of 6 hours in the city this month, which is about all of the city I can stand these days.

I love working from home. Seriously, I've been trying to figure out how to do it for years, and now here I am. I've got a MacBook Pro, a somewhat reliable internet connection and a license for Adobe Creative Suite... what else could a person need?

My internet connection here in rural Virginia is a little box manufactured by NetGear. It connects to the Sprint tower over by the fire station through the magic of 3G and, occasionally, when the stars are in alignment, LTE. That tower is connected via underground cables to the whole, wide world. I sent a proof to my friend John Horne in the UK this afternoon. Magic.

I hope everyone has a great time at PTC. Don't forget to have a mai tai for me, and a glass of fresh-squeezed pine-

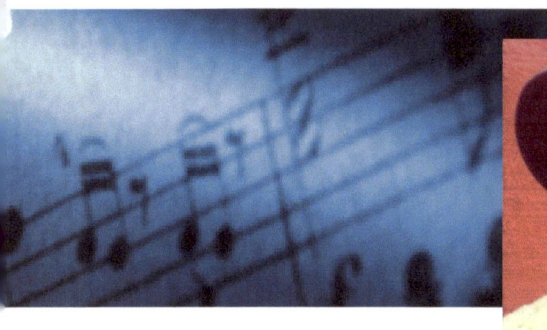

apple juice at the Royal Hawaiian. I'm going to be here on the farm, milking cows and feeding the hogs. And if you need me, I'm just an email away.

Kevin G. Summers is the Editor of Submarine Telecoms Forum and has been supporting the submarine fibre optic cable industry in various roles since 2007. Outside of the office, he is an author of fiction whose works include <u>ISOLATION WARD 4</u>, <u>LEGENDARIUM</u> and <u>THE MAN WHO SHOT JOHN WILKES BOOTH</u>.

 +1.703.468.0554

 <u>editor@subtelforum.com</u>

www.ingramcontent.com/pod-product-compliance
Lightning Source LLC
Chambersburg PA
CBHW041204180526
45172CB00006B/1183